Telecommunications
for Learning

The Educational Technology Anthology Series

Volume Three

Telecommunications for Learning

EDUCATIONAL TECHNOLOGY PUBLICATIONS
ENGLEWOOD CLIFFS, NEW JERSEY 07632

Library of Congress Cataloging-in-Publication Data

Telecommunications for learning.
    p.   cm. — (The Educational technology anthology series ; v.
3)
    Articles reprinted from Educational technology magazine.
    Includes bibliographical references and index.
    ISBN 0-87778-225-3
    1. Telecommunication in education. 2. Distance education.
I. Educational technology magazine. II. Series.
LB1044.84.T45   1990   90-41902
371.3'078—dc20   CIP

Printed in the United States of America.

Library of Congress Catalog Card Number:
90-41902.

International Standard Book Number:
0-87778-225-3.

First Printing: January, 1991.

Table of Contents

All of these articles are reprinted from *Educational Technology* Magazine, published by Educational Technology Publications, Englewood Cliffs, New Jersey 07632. This volume is one of a series of anthologies of articles from *Educational Technology*; for original dates of publication, see pages 189-190.

Telecommunications
for Learning

Part I

Focus and Setting

The New Age of Telecommunication: Setting the Context for Education

Dan J. Wedemeyer

The purpose of this article is to provide an overview and the technological context for the new age of telecommunication. I will discuss the basic systems and services now offered or which will be offered in the near future. The main focus here is to aid in the overall development of what I call *integrated telecommunication literacy*. That is to say, no longer can we afford to view ourselves as single-focus, stand-alone educators, administrators, curriculum experts or media specialists. We can no longer limit ourselves to *any* single perspective. Telecommunication technology and service integrates our concerns and calls for new, innovative approaches to learning and living. Our jobs have changed, our tools have changed, and our clientele's needs have changed.

It is generally accepted that to get along in the world today one needs to be "literate." This has commonly been referred to as the ability to read and write. Because of developments on several technological fronts and the convergence (networking) of these major technologies, we live in a world that is increasingly dependent on forms of communication that were never before available to humans. At this moment, sophisticated and increasingly-integrated telecommunication systems and services are being used by governments, high schools, universities, corporations, and individuals. Soon, those of us *unable* to effectively and efficiently use these additional tools and processes of interaction, expression and learning will be as "functionally illiterate" as the person in the past who could neither read nor write. As educational professionals, we have a responsibility to anticipate the new skills requirements and develop new literacies. The need for this new form of literacy

Dan J. Wedemeyer is Chair, Graduate Field of Study, Department of Communication, University of Hawaii at Manoa, Honolulu, Hawaii. This article is adapted by the author from his Keynote Address at the 1986 National Technology Leadership Conference.

is a pressing national educational problem, not only for us as professionals and our student clients, but also for the wider public, as life-long learning becomes a widespread priority.

There is no single telecommunication technological development that has caused the changes that we are experiencing today. Rather, it is the convergence or *integration* of several technologies that is creating new possibilities and new patterns of communication and learning. Many point to the development of the microchip as the driving technology. Others see the use of satellites for global communication as the driving force. In fact, these and other technologies (lasers, optical fibers, intelligent robotics, etc.) are principal characters in the transformed telecommunication-based society. Some elements certainly play larger roles, but it is the "synergy" or networking of these developments that has brought about today's global communication environment. Let's look at some of the developments which are driving forces in this new environment—specifically the telecommunication systems and services which are in various stages of development today.

Telephone Technology

The telephone is the standard instrument of modern communications around the world. It has been with us for a little over a century and a sophisticated global network has evolved. Some have called the global telephone network the largest machine in the world. Initially intended for voice communication, the telephone system is increasingly being used for data (including data over voice applications), compressed video, facsimile, electronic bulletin boards, telesketch pads, and other types of electronic transmission. The latest developments in distance learning often involve one-way video out and feedback links via 800 numbers. Telephone lines and modems often serve local PC networks for drills and examination function. Existing telephone systems especially in hybrid networks will play a major role for several decades even though faster and greater capacity communication systems are becoming necessary and available. Present telephone networks provide an in-place, *two-way* communication link between hundreds of millions of users. The telephone companies have a sizable capital investment in the paired copper wires, satellites, cables, and microwaves connecting their users.

However, the traditional telephone network is too slow for data communication because it is an *analog-based* system. Data communication and increasingly all forms of telecommunication is in a *digital* form. Conversions between the analog and digital signals are necessary to create the integrated continuous link using today's networks. Transformation of these analog networks seems to be an expensive necessity, and larger-capacity, all digital, "smart" networks are on the horizon.

Cellular Telephone Technology

Cellular Mobile Radio Telecommunication Systems are an emerging technology. They are also referred to as CMRTS, Cellular Radiotelephone, Cellular Radio, Cellular Telephony, and Cellular Mobile. Cellular Telephone offers the next step in liberating the user from the umbilical wire while staying on the network. Mobile radio has been a reality since the early part of this century with ship-to-shore communications. Mobile telephony has existed for a number of years but its quality and availability have been low and the cost high. The problems of traditional land-mobile telephony have to do with limited over-the-air *frequencies*. There are simply not enough to service all potential users.

In other words, the problem with mobile telephony has been scarcity of available *channels*. Enter Cellular Telephone. Cellular Mobile telephone divides a coverage area into small sub-sections or cells. By using low-powered transmissions, the same frequency can be reused several times as long as each use is in geographically separate cells. Utilizing computerized switches, a mobile telephone call can be transferred ("handed off") from cell to cell as the caller travels through the coverage area. Switching occurs so rapidly that it is unnoticed by the human users of the system. The quality of the signal remains consistently clear. A centralized switching unit ties mobile users into the local telephone company and the global networks.

Broadcasting Techniques

Telephony, whether by cellular mobile or by wire pairs, is point-to-point communication, usually one-to-one. Broadcast radio and television, on the other hand, is a communication from one-to-many. It is an efficient and relatively cheap-per-capita way to *send* messages. It has been used widely to decentralize education. Now satellites are increasing coverage footprints. Governments have found the broadcast media an effective way to inform and educate the public, set development agendas, disseminate propaganda, and to maintain centralized control. Because of its mass appeal and high visibility, broadcasting has been considered the dominant form of electronic communication, but it too is changing. While in the near future it will no doubt retain this position, in the long term other forms of telecommunication may erode this power or be integrated with this highly centralized, one-to-many communication system. Video, and increasingly digital video, must become a two-way medium. The future broadcasting system will be more interactive, more user-controlled, and all-digital in nature.

In the long term, lack of user control may be the single most negative aspect of traditional broadcast communications. Already, videocassettes and videocassette networks allow more *user* control and less centralized message management—other alternatives are forthcoming.

In addition to delivering video messages via tape, satellites have been utilized by broadcasters for two decades for domestic and international distribution. In the early days of satellite communication, the earth stations used to receive and transmit were expensive and often required an engineer's constant care. Today, improvements in the transmitting power of Ku-Band satellites now makes it possible to broadcast directly from satellite to home via reception on a roof-mounted dish about one meter in diameter. This makes television reception possible in even the most remote areas—this system is known as Direct Broadcast Satellite (DBS). Systems using slightly larger dishes (three or four meters in diameter) for television reception in rural areas or areas outside primary footprints are operational or have been successfully tested in many locations.

Cable Television Technology

Alternative delivery forms for broadcast television also exist. Cable television, for example, already integrates the capabilities of broadcast TV and builds expanded capabilities for other services. Indeed, it may become an information utility on par with electrical or water utilities.

Cable television first began as a system to provide television to communities unable to receive a broadcast signal. Early cable systems were one-way. As communication satellite use increased and transponder cost and receiving dishes became more competitive, the opportunity for widespread distribution of specialized programming (sports, movies, religion, news) became possible. Educational services have not, as of yet, utilized network cable TV on a truly national scale; selected institutions have taken advantage of the medium.

Now, interactive cable allows the subscriber greater freedom of choice in programs and a variety of added features and services. For instance, home security for burglary, fire, flood, and other threats can be handled through the two-way system. Home shopping, interactive videogames (with a central computer or other systems subscribers), conferencing, videotex delivery, interactive instruction, and at-home banking are a few of the *possibilities* of such systems.

Multivision in Japan incorporates cable TV and videodiscs for educational applications. The market is still exploring, indeed creating its potential. But it holds one of the keys to interactive educational services.

Fiber Optic Technology

Beyond coaxial-cable-based television is the broadband information utility possible through fiber optics. Fiber optics communication, the process of transmitting information by light waves through thin silicon (glass) strands, is one of the integrated delivery options cablecasters, telephone companies and designers of data networks are considering building. Optics provide a way to greatly increase the capacity of their systems, e.g., optical fibers are extremely thin (about 125 microns in diameter) and lightweight, yet have far greater capacity for circuits than any metal cable. Recently, AT&T Bell Laboratories set the world record for transmission capacity of a lightwave communications system—20 billion pulses of light per second. The equivalent of 300,000 voice conversations, sent 42 miles, on a hair-thin fiber of supertransparent glass. This certainly has ramifications for educational applications. Tradeoffs in size allow more space in crowded conduits now under the streets. They are also planned as transoceanic transmission systems in the near future. AT&T is going to build the first lightwave communications systems under the Atlantic Ocean, and a similar system is planned for the Pacific. In 1988, laser beams traveling through two pairs of glass fibers will carry the equivalent of 37,800 simultaneous conversations overseas, under water, from the U.S. to Europe and the Far East. The routes for this system total 21,000 miles. Because the signal travels by light rather than electricity, it is not subject to distortions from electromagnetic interference that plague wires, nor is it easily tapped. These new cables are international electronic expressways of the future.

An added benefit of optical fibers is the low requirement for amplification. Electrons moving through a wire encounter resistance and periodically the signal must be *regenerated*. Essentially this means giving it a boost or push to keep it moving down the line. Signal regeneration is required about every eighty miles for optical fibers compared to every mile for conventional cables. This presents a considerable savings, especially in long distance lines.

The light that travels through the fibers is supplied by lasers which emit a concentrated beam. Fibers can handle television, data, and voice information with equal ease. The twisted pair (wire) in traditional telephone systems provides a *narrowband* channel which can handle voice and slower data transmission rates. A new fiber optic system recently put into operation and originally developed by Bell Labs can transmit one billion bits of information (the equivalent of 20 digital TV channels, 14,000 telephone conversations, or 100 average novels) each second.

Fiber optics may provide an attractive alternative to satellite communication for some applications. The inherent characteristics of satellites will most probably guarantee a central role of these space-based systems, but thick-route communication fibers appear to have the future edge for truly integrated services.

Satellite Technology

Communication satellites connect the most remote spots of the world. The circuit capacity of each new generation of satellites has increased as the cost of transponders (combination of receiver, converter, and transmitter) has gone down. Satellites can be used for all types of information—voice, data and video. There are international systems (Intelsat), numerous domestic systems (RCA, Western Union, AT&T, etc.), and private systems (Satellite Business Systems), and a host of educational-based satellite networks. Earth stations can be fixed or mobile. Newer highly-powered satellites no longer require high-cost, high-maintenance receiving dishes. Ku-Band, Direct Broadcast Satellites (DBS), for example, can transmit directly to individual homes. A combination of satellite, terrestrial-microwave links and broadcasting radio frequencies may soon provide a national paging service in the United States.

A new generation of larger communications satellites is now possible. Joseph Sivo, chief of the Lewis Space Communications Division, predicts, "Technologies to be tested . . . could lead to at least a five-fold increase in satellite communications capabilities in the 1990s (to 180,000 circuits). These capacity increases will be necessary to meet the rapid expansion of telephone, television, teleconferencing, electronic mail, data communications, and other communications satellite traffic for the rest of this century—growing at roughly 20 percent per year."

The next step may be the construction of satellites in space on orbiting space stations and the clustering of a group of satellites to a platform equipped with broad switching capabilities and intelligence. Both will lead to further cost reductions and increased efficiency.

Computer Technology

Another key telecommunication technology is, of course, the computer. Interesting developments abound in the areas of the super computer, personal computer, and the latest development, the credit-card sized "smart card."

The story of computers is really the story of the *silicon integrated circuit*. Today, a computer is essentially a non-intelligent machine that is capable of performing and storing millions of operations in a short time. In 1959 all of this was produced on a single silicon chip for the first time—the *integrated circuit*. Size and power consumption again took a quantum leap toward miniaturization. Had it been available in 1959, one megabyte of memory would have occupied about 400 cubic feet—a room seven feet square and eight feet high. Using 256K chips, it now would require one-half cubic inch of space. In just the past 10 years, the cost of *this* memory has gone from $600,000 to below $1000 on a microcomputer. Size and cost reduction have increased terminal penetration from one-per-seven white-collar workers in 1980 to one-in-three in 1985. The trend will no doubt continue.

Since 1959, the emphasis has been on placing more and more circuits on a single chip—from ten components on a chip in 1964 to one megabyte chips (1 million components on a wafer-thin chip) in 1985. At least one supplier (AT&T) of the one megabyte chip claims that access to the information on the chip is 20 million bits per second. That means it can supply 1,048,576 bits in the time it takes to say "bytes."

Looking toward the year 2000, today's technology predicts sub-micron design widths, with lines only 400 atoms wide. Chips containing 100 million components—50 times the current amount—could have tiny *regions* with capabilities that more than match the megabit chip, and perhaps by the turn of the century a billion components can be placed on a chip about the size of today's postage stamp. I might add here that such a chip at that time may cost a lot less than the postage one might need to mail a letter in the year 2000 . . . if indeed we are still mailing that way!

Work is progressing in Japan, Europe, and the United States on the so called "Fifth Generation Computer" with 100 times the power of the largest computers available today. Japan's Superspeed Project is a $200 million effort to produce by 1989 a supercomputer that performs a billion operations per second. Intense work is ongoing to develop a new computer that may be the first truly intelligent machine—able to "reason out" solutions to problems for itself or in a partnership with humans. They will be "knowledge processors" as opposed to "data processors."

Movements are underway to develop expert systems, the base of AI (artificial intelligence). Undoubtedly we will soon have computers we can talk with or that will translate foreign languages into our native tongue in real time.

Generally, computers have been placed in four categories, according to computational ability, physical size, and cost. The so-called supercomputer, of which fewer than 150 exist in the world, are the largest, fastest, and most expensive. Cray Research has introduced the world's most powerful supercomputer, the Cray-2. The new machine, employing 240,000 chips immersed in liquid coolant, will deliver one billion calculations a second while sending and receiving data from 36 disk drives simultaneously. It will cost about $17 million each.

Cray says the new machine is significantly faster than the other supercomputers now in existence (97 of which were made by Cray). The first Cray-2 has already been installed at the Lawrence Livermore National Laboratory, in Livermore, California, where it was assigned to unravel some of the fearsomely complex equations governing nuclear fusion. The second category is referred to as mainframe computers. They are less expensive ($20,000 to $400,000 U.S.) and used primarily by large military and government institutions. Convex Computer (Dallas) hopes to bring many of the capabilities of supercomputers to a wider market with its $500,000 C-1 machine. Based on the Cray 1's architecture, but with only one-fourth its power, the C-1 could bridge the large performance gap between supercomputers and superminicomputers such as Digital Equipment's VAX line. The third category, microcomputers, are considerably cheaper ($200 to $5,000). Minicomputers, the fourth ($20,000), fills the gap between the two. Now a fifth size computer, sometimes referred to as a "smart card," places the processing power and storage of a microcomputer in a plastic envelope the size of a credit card. This card is truly a personal computer and can be tailored to a single individual's needs. The true microcomputer card houses an 8K memory and an 8-bit processor and costs from as low as $25 up to $100.

Increasingly, networking and computer communication allows the individual computer owner with a modem to tap into the power of mainframe computers, remote data bases, and other microcomputers.

Interfaces between different languages are becoming more common and a standard language may not be far off. We are at the beginning of an exciting telecommunication era in which almost all electronic devices will contain some form of computer. "Smart" learning stations, "smart" homes, "smart" phones, etc. While processing power of these computers has gained dramatically, storage has been a problem. Substantial gains in storage technologies have recently been realized in the videodisc and compact disc.

Optical Disc Technology

Videodisc technology has demonstrated vast potential, which has been largely unrealized in its application. Videodisc provides a method of information storage and retrieval that far surpasses existing systems. Combined with a microcomputer and/or cable TV, it gives a dramatic new meaning to interactive visual teaching systems. Generic discs are being developed that can be "programmed" with a wide range of applications based upon user needs. Rapid random access to any frame of the laser-read disc assist in computer-managed applications. More than fifty thousand separate frames can be permanently stored on one side of a disc. New techniques referred to as WORM (write once, read many times) and DRAW (direct read after write) will replace many of the educational applications of pre-recorded discs. Erasable discs are on the horizon. Hitachi could soon lead this industry by producing erasable laser discs using a color memory alloy recently developed.

The immense capacity for storage and random, fast retrieval makes the laser disc an ideal instructional and reference tool. Networked with a microcomputer, vast amounts of visual information or data can be scanned, reviewed, combined, and managed to provide many new services.

The much-discussed technology of optical disc will undoubtedly have a significant impact upon education as well as existing data bases and other repositories for information.

In the future, coupled with the personal computer, videodiscs and the newer compact discs (CD ROM) promise a highly human-factored solution to information storage and education problems.

Telecommunication Services

The previous sections have described briefly the major technical systems of the transformed telecommunication society. They are *the means*, the conduits, the tools at our disposal. Now attention is turned to the services, existing and emerging, available via these systems. The services are not confined to a single system; rather, they utilize the best combination of technical means for delivering an information service. In many cases the need for particular services created the technologies and the networks.

In the longer term, all the services I will describe will be forms of data communication. They most

certainly will be all-digital within the timeframe mentioned earlier in this presentation. Data/digital communication requires fast, accurate transmission capabilities. More and more data communication services are being offered with digital switching and all-digital transmission paths. Indeed, in the future, the *integration* of services will be accomplished through high capacity, all digital, *smart networks*. A good example of this is seen in the latest generation of office automation.

Office Automation Services. Office automation is providing more options for the way information is handled in business or educational settings by consolidating some functions and decentralizing others. Organizations best-suited for telecommuting have a healthy proportion of information-type functions.

In the central office and in home offices, typewriters have turned into word processors, and microprocessors are adding sophisticated word processing and message handling software. Copying machines are now linked into facsimile networks. The future seems bright, indeed, for the tele-facsimile market. Arthur D. Little predicts that by 1990, the cost per page will drop below the price of a first class postage stamp, and believes that it is only a matter of time before facsimile copy quality will be on par with that of office (or library) copiers.

International Resource Development forecasts that the overall facsimile market is projected to grow from shipments of 75,000 units worth $346 million in 1983 to 212,000 units worth $827 million in 1992. The average price per unit is expected to decline from $4,600 in 1983 to $3,900 in 1992.

The microcomputer is becoming as standard in the office as file cabinets. The "smart desk" is often mentioned as the standard work or learning station of the future. Data communication networks (sometimes accomplished via Local Area Networks) provide the latest information to aid in those learning or decision situations; and, just as quickly by electronic or voice mail, make students or associates aware of their actions. In a report called "Electronic Mail Systems," LINK Resources estimates that overall electronic mail service revenues will grow from $80 million at year end 1983 to more than $2.1 billion in 1988. Computer-based messaging service revenues are expected to grow from $122.5 million at the end of 1984, to $1.3 billion in 1988 for an average annual growth rate in excess of 100 percent.

Further, voice-mail revenues, for instance, will climb from $214 million this year to more than one-and-a-half billion dollars in 1991. With a home terminal and modem, selected staff members or students can accomplish their tasks without traveling to a central office or campus. Professionals can stay up-to-date while traveling by using a portable terminal that fits in a briefcase and plugs into a standard phone jack. It appears that an option of at-home work and learning will evolve.

Electronic Funds Transfer Services. Electronic Funds Transfer Services or EFTS marks the worldwide movement away from a paper and coin medium of exchange to an electronic one. Money in many situations has been transmuted into information. The recent spread of the Automatic Teller Machine (ATM) has brought the concept to the public. In the U.S., banks now perform more customer transactions at automated teller machines than in human-staffed lobbies. ATM service provides 24-hour, seven-days-a-week access to checking and savings accounts. Customers insert a plastic computer car or smart card in the ATM and enter a personal code (PIN) to interact with the system. At-home banking packages are being offered via telephone, personal computers, and videotex.

Videotex Services. *Videotex* is the generic term referring to electronic message of text and graphic material. There are many terms used to refer to this emerging information service, causing more confusion than clarity. Essentially, there are two types of videotex—one-way and two-way. One-way delivery is known as *teletex*. It is also referred to as *broadcast videotex*. Two-way delivery is known as *videotex*, videotext, view data, and interactive videotex. Videotex is very similar to online database services. In fact, it has been called an online service with color graphics. The use of *graphics* is an identifying characteristic of videotex. It has two other primary characteristics. It is a *universal service* and it is *user-friendly*.

Teletex is usually delivered to the home via the *vertical blanking interval* on a standard broadcast or cable television channel. Teletex may also be sent by wire or cable.

Videotex services delivered by cable are usually interactive. The cable provides a two-way path. However, the most common medium for interactive videotex is the telephone system.

There are also hybrid delivery options. The pages of information may be delivered via broadcast teletext with subscribers connected back to the system by a telephone line. Once the signal reaches the user's premises, regardless of the channel, it must be converted into a video display, or be used to activate a video disc for full-motion television.

Teletext services usually concentrate on rapid delivery of information that is constantly changing—so-called perishable information. Interactive videotex offers unlimited databases as well as many two-way services such as banking, computer-aided

instruction, on-line shopping, electronic mail and so forth. Teletex can reach more people and is an inexpensive service. Interactive videotex serves fewer people at one time and is usually more expensive than teletex.

The importance of videotex in the transformed telecommunication-based society is its potential to provide a universal information service (an information utility). Videotex promises to eventually spread the information revolution to the general public, but the business or education market seems much more viable.

The future of videotex promises more integration. Terminals may merge with personal computers, creating an integrated home telecommunication appliance. Use of videodiscs for storage and optical fibers for delivery may provide high speed, high quality transmission of photographic-quality graphics.

Teleconferencing Services. Perhaps one of the most promising interactive services is teleconferencing. Teleconferencing is an electronic meeting between more than two people. The three basic types of teleconferencing are *video, audio,* and *computer.* An education-oriented teleconference may include any combination of these types.

The simplest type of teleconference is a conference telephone call. The most complex would be computer-assisted, full-motion video with multiple audio channels and facsimile.

In the past teleconferencing was limited to small groups and was expensive. People had to go where the technology was (TV studios, conference centers, etc.) and it took a great deal of preplanning to set up the network connecting participants. As the hardware has become more mobile, personal, and less expensive, teleconferencing has diffused to many user groups. This trend is expected to continue especially in training applications.

One of the most important benefits of this class of services is its substitutability with travel. Teleconferencing saves transportation inconvenience, costs, and time. Teleconferencing is not replacing face-to-face interaction, but is greatly supplementing it. Communities of interest, connected electronically, may be as common as traditional geographic communities are today.

Future Trends

What can we expect in the years to come? For the immediate future three trends will continue. That is, telecommunication miniaturization, digital integration, and networking efficiency. Storage devices and processors will continue to shrink in size and cost and increase in capacities. Even 128K RAM standards for many of today's home computers are already obsolete and insufficient indeed with truly human-friendly software requiring large temporary storage areas. In the next five years, a new generation of personal computers will offer users document and graphic design, tools, artificial intelligence software, and advanced professional assistance. By the year 2000, personal computers will have eight billion bits of main memory, and they will be affordable by almost anyone.

The next key requirements will be for operating system compatibility and software portability. Unix seems to be addressing many important elements in this regard. Developments at GM and Boeing in their current activities are also promising as each is moving to "standardize" office and manufacturing communication protocols. Integration will continue as the hardware merges into multi-use units. It is already common for word processors and modems to be standard features of computers rather than stand-alone units. The telephone is evolving from a simple dialing and voice translator into a smart data device. Indeed, the personal computer will probably be the phone of the future. Until then, the touchtone telephone will probably become standard as its use expands beyond merely entering phone numbers to acting as a terminal input/output device for your memocard. Eventually the telephone, computer video-tex terminal, videodisc, radio, and television set may merge into one home (modularized) appliance. High Definition Television (HDTV) is likely to provide motion picture-quality images for all video services, including videoconferencing.

The most significant development in integration, however, is not in terminal equipment but the system of delivery itself. Currently a number of paths and conversations must be used to connect many types of equipment. In the next few years more and more will be heard about the *Integrated Services Digital Network* (ISDN). This is a network with two main attributes that enables it to service all telecommunication needs. It is *universal* and *intelligent*.

ISDN may not reach homes for many years. The current telephone wires that connect each subscriber are not suitable for ISDN. It will be a long process to replace these with broadband carriers, probably light fibers. Once this is accomplished, we will have an information/education utility going into each home.

What all of this will finally mean is increased efficiency and flexibility in telecommunications on a scale that can scarcely be imagined. Education applications will need to take advantage of this new environment and train people to be literate within it. □

Educational Technology Use in Distance Education: Historical Review and Future Trends

Robert A. Gray

The day was extremely cold. A petite young woman was relaxing in a comfortable chair in front of her fireplace with a notebook in one hand and a hot drink in the other while attentively watching television. She was a student in a distance education program. Evans (1986) has defined distance education as the delivery of credit and non-credit instruction where the majority of content expertise and management is at one location and the majority of student learning activities at another. Feasley (1982) has defined the concept simply as learning that takes place at a site remote from the instructor. Correspondence courses, television classes, and interactive systems via the computer are all examples of distance education programs.

The purpose of this article is to examine the potential of educational technology utilization in distance education. Specific examples will be examined with particular emphasis on the British Open University (BOU), a pioneer in the development of distance learning projects. A review of technology-related delivery systems used in distance education will be critically analyzed. The last section of the article will examine trends in the use of technology in distance programs.

British Open University

In 1969 the British Open University was chartered as a degree-granting institution serving working adults who could not attend the regular British universities. Seeff (1981) described the BOU in concert with the BBC, British Broadcasting Corporation, as one of the stunning successes in higher education.

The BOU is a unique example of an innovative learning system. The characteristics which distinguish it from traditional programs were suc-

Robert A. Gray is Associate Professor of Educational Media and Technology, East Texas State University, Commerce, Texas.

cinctly described in the following manner (McIntosh and Wooley, 1980):

- The program was designed primarily for working adults.
- It was mainly home-based and part-time.
- For admission, there were no educational prerequisites.
- It used a wide variety of media and print materials.

Course work at BOU was planned so that two-thirds of the students' study time could be allocated for systematic reading, fifteen percent for face-to-face contact with staff and other students, ten percent for viewing broadcasts, and ten percent for doing written assignments and taking exams. Although texts and broadcasts were designed to assist students to learn independently, tutors were also available. The tutors were usually staff members from conventional universities who desired to earn additional money, were retired faculty, or were women with advanced degrees who preferred part-time employment (Stevenson, 1983).

Although Open University students primarily did their assignments at their residence, they could also use study centers, a support service provided by the program. Every student was assigned a center which was located in populated areas accessible by mass transit. The centers were used as meeting places for tutors and students. This association was aimed primarily at promoting scholastic attainment; however, tutors were also most interested in providing individual guidance to BOU students. Correspondence units and broadcasts provided students with basic content in a fairly conventional manner. However, the tutorial and counseling divisions interpreted the program in relationship to the individual needs of each student (King, 1982). The play and subsequent film, "Educating Rita," described in a humorous manner the relationship between a working-class woman and her tutor-counselor. The counseling and tutorial functions of tutors were dramatically emphasized in this film.

The courses at BOU utilized a mix of print and media resources. The principal print resource was the study guide or text, a compilation of lecture notes using an interactive design. The guide took the place of formal classroom lectures. BOU also used audio cassettes, records, television, films, kits, and depending upon the specific course, maybe even a microcomputer or a laser disc. Television was used extensively for enrichment and to illustrate concepts; it was not designed to replace the print media (Seeff, 1981). Students in most classes were assigned weekly half hour television programs to watch and/or to listen to radio programs. In addition, students were sent broadcast notes that related the program to course content. The administration also provided audio and videocassettes for individual use. With the growth of videocassette players in the home, this method of instructional delivery should expand in importance.

A unique feature of the program at BOU was the team approach used to design individual courses. In fact, subject matter specialists collaborated with educational technologists to design, produce, and revise the numerous courses involved in the curriculum. A large staff of specialists was located in Milton Keynes, England, to perform these tasks. The subject matter specialist provided appropriate content for each course design as well as assistance in developing other components. The educational technologists supplied expertise in defining course objectives, specified teaching strategies, planned learning experiences, and developed evaluation plans. The team approach represented a positive element in course design for the BOU program.

Today, the British Open University has fallen on difficult times. A tight budget and a socialistic image is plaguing the school (Walker, 1984). In 1984 these problems resulted in a major decrease in enrollment of 3,000, to a total of 22,000 students. In addition, the University is changing curriculum focus from providing purely undergraduate programs to the development of a more comprehensive offering including vocational courses. All of these new developments will effect the use of technology at BOU during this transitional period.

Technology Utilization

Definition

In this article educational technology has essentially three distinct meanings. The intended meaning can be readily deciphered by analyzing the context. Heinich (1986) defined technology in three distinct aspects. Technology can be a product, a process, and a mix of both product and process. For example, as a process it represents the systematic application of scientific knowledge to education. Educational technology involves the process of systematically designing, implementing, and evaluating the total teaching and learning system. Technology as a product refers to the hardware and software used in instructional systems. Technology is also often defined as a mix of product and process. Distance education utilizes both aspects of this definition in the development of curriculum.

Beyond definitions, distance educational specialists must be concerned with other crucial issues concerning technology use. For example, admin-istrators of these programs should be producing cost-effective curricula utilizing technology-related delivery systems while maintaining positive results. A review of the wide variety of educational technology for distance education provides planners with many choices. However, curriculum developers should realize, according to many media theorists, that there is no research basis in terms of learner outcomes for selecting one medium over another.

Print

In most distance education programs, print materials in various forms are widely used. The format is relatively inexpensive, highly portable, and can be transported readily through conventional mail services.

Correspondence study using existing mail systems has been in existence for many years and because of obvious advantages will continue. Telecommunication delivery modes, including electronic mail and computer use, could facilitate the process by decreasing time of message delivery, increasing the degree of interaction by students, and, perhaps, even result in greater student interest.

Television

Bates (1979) described several functions of television and radio use which are very general and apply to most of the programs at BOU. Five specific items are included in the following section:

1. To increase the students' sense of belonging; to make teaching less impersonal.
2. To reduce the time required by students to master content.
3. To pace students; to keep them working on a regular schedule.
4. To attract new students to the programs or to individual courses. In addition, to interest general viewers in subject matter.
5. To establish academic credibility of courses and programs to the outside world.

A variety of different types of delivery systems can be used with the television medium. Open circuit public television was a popular format, followed by cable and commercial TV. With the decreasing price of videocassette players, many students will have access to this new medium. The relatively small videocassette can be transported easily, even using mail services. In addition, the student can fast-forward through the tape to locate specific segments for further study as well as to repeat important concepts. Substantial growth in videocassette use in distance education programs in the future is very predictable.

Television as a delivery system has pronounced advantages and noteworthy limitations. The medium can reach large audiences with distinct mass appeal. Television viewing has become so popular that it consumes six hours per day in the average home. Senior citizens, preschoolers, and other segments of the population often watch TV beyond the six hour statistic. Financially, the cost per pupil also can be relatively low because of the large potential audience. Lastly, reams of paper describing research studies are available documenting the effectiveness of television as an educational tool. However, limiting the impact of instructional TV, some studios, even the comprehensive facilities, sometimes have produced substandard programming. Ineffectually designed programs frequently have not taken advantage of the uniqueness of the television medium. Programming which primarily uses dull, "talking heads" as the primary instructional strategy has neglected a host of additional stimulating techniques. Adverse atmospheric conditions, power failures, and equipment malfunctions, obviously, also have influenced TV viewing. Moreover, most TV systems are designed to be linear, a one-way communication system. By means of telephones, computers, and print correspondence, distance education projects are now including feedback loops as a means of minimizing the negative effects of one-way transmission.

In emerging countries television and radio over high tech systems seem to be increasing in popularity. Countries with vast areas of barren land and a meager population base are forced into selecting a telecommunication delivery system for economic reasons. For example, in Western Australia schools have been using television and radio for many years with great success. China also has been using telecommunication systems, television and radio, to educate the vast number of students that can't be accommodated in the traditional universities. Since 1979, 240,000 people in China have graduated from distance learning programs (Turner, 1986). With the large population and the vast geographical area, the Chinese found that the most cost-efficient delivery medium to meet their needs is a telecommunication system.

Cable

Cable television can produce positive learning outcomes within a cost-effective context in distance learning programs which are located in close geographical boundaries. The former Warner-Amex "QUBE" is an excellent example of a cable program which used a two-way transmission. Other cable formats obtained feedback by using specific classroom sites or by extensive utilization of the telephone. QUBE used a keypad at-tached to the system to input specific responses. For example, students enrolled in a literature course could relay their answers to questions asked by the teacher simply pressing a single key on the keyboard.

Cable utilization is rapidly expanding. A high percentage of households in the U.S. are serviced by cable companies providing a multitude of channels from pay movies to religious programming. In the future more homes will have cable with far more channels—up to even 1000 is possible. With this mushrooming popularity, distance educators need to capitalize on the educational potential of the new format. Well-designed instructional programs could be broadcast through cable systems utilizing colorful graphics and with a feedback system to provide a stimulating learning environment. With the increase in channels, costs would probably decrease, thus resulting in a more cost-effective system. Cable television has tremendous potential in distance learning programs in local geographical areas; to broadcast courses over a large land mass, broadcast television or radio would prove to be more cost-effective.

Videodisc

The videodisc has some unique advantages as well as important limitations. Because of its many features, the videodisc can be designed to be extremely interactive, leading to a positive learning experience. The major negative dimensions of videodisc technology relate to the lack of record capability and high production costs (Gray, 1978).

Satellites

Satellites have been used to deliver audio and video signals to sizable audiences covering vast geographical areas with much success. The large, widely-dispersed audience helps to make this medium cost-effective. For example, India was covered by satellite transmission, broadcasting educational programming using U.S. technology. The many dialects in India created major software design problems and subsequent headaches for producers. There are other examples of instructional television via satellite in this country. College courses were distributed over satellite by the National University Consortium and the Appalachian Education Satellite Network. The National University Consortium represented 66 universities, including the Smithsonian, which shared noncredit courses over satellite transmissions. The Appalachian program used a satellite delivery system to broadcast staff development education courses to teachers in remote mountain areas. In a few years, 1,200 teachers received grade credit from 13 institutions. Courses were provided 64 hours a week at

20 community sites and 220 cable systems (Lewis, 1983). With hardware systems in place, distance learning programs ought to investigate satellite transmission of programs specifically if the programming area is extremely large. With decreasing costs of technology, satellite use in extended educational programs will surely increase in popularity.

Emerging Technologies: Teletext and Videotex

These are two new information services that utilize special TV receivers. Teletext programs involve a one-way transmission of information, while videotex uses telephone lines to communicate in a two-way interactive system. A common use of teletext is found in cable systems that present weather and news data in printed fashion. A variety of new teletext and videotex systems currently are being tested in France, Japan, Britain, and the United States. The development and application of these new systems is occurring primarily in the industrial training field. The potential for use in distance learning will be more clearly defined as findings from industrial use become known.

A number of uses for teletext and videotex systems for distance education can be examined. For example, catalogs, syllabi, assignment papers, etc., can be quickly transmitted to the learner; feedback can be obtained rapidly in an interactive format; registration and course selection can occur; and timely information can be communicated to the students. Combined with a cable system these emerging technologies should be cost-effective.

Computers

The mighty computer also has been used effectively in distance learning programs. For example, in Canada the Telidon System combined with a teleconferencing hook-up has been used successfully in distance learning programs for some years (Winn, 1986). The project has used a two-way system to present print and graphics to teachers enrolled in grammar and classroom management courses. In Japan computer use is focused on research activities rather than educational applications. Students in Japan rarely use computers until they enter industry where training programs utilize computer delivery systems. Computer-related programs in distance education will find more uses for the application of this delivery system as findings from successful programs are known. Growth in computer use for nontraditional and traditional education is inevitable.

In addition, microcomputer faculty use depends upon a number of important issues. First, a strong support system must be present to assist faculty.

The development of the program based upon a needs assessment; appointment of a coordinator; the conduct of staff development computer workshops using a humanistic model; the purchase of hardware and software in relationship to needs; and ready access to resources all represent major problem areas that distance educators, in fact all educators, planning to use the micro need to examine (Gray, 1984). With the tremendous power of the computer, distance education programs can quickly process all types of alphanumeric data, thereby increasing efficiency. In the future, growth in the application of micros to extended learning will continue at a fairly rapid rate.

Radio

Distance learning programs have been using radio as a delivery system throughout the world for many years. The experience of the BOU highlighted the major advantages of the medium. Bates (1979) indicated that in order to maximize learning students ought to be taught the role of the different types of radio programming. For example, some programs are essentially a lecture while others are primarily documentaries. Students learn from both programs, but with a preference for documentaries by curriculum designers.

Today, radio can be used very effectively for distance education. For example, course content can be broadcast over large distances to a sizable audience. It usually takes the format of a short lecture, a drama, a poetry reading, a guest presentation, or a question-answer session utilizing the telephone. In addition, radio can be used to reinforce content disseminated through the print medium.

Summary

Feasley (1983) noted several trends related to media use in distance learning programs. He described the following:

1. Two-way systems are becoming more popular by the addition of an interactive component to the traditional one-way systems. The trend is toward interactive systems and away from broadcast delivery media.

2. Interactive systems will probably cost more, and will represent a delivery mode that will be more difficult to share with other schools. It is anticipated that the lack of sharing of delivery software will precipitate the escalation of expenses for nontraditional, extended learning programs.

3. The more complex delivery systems will lead to less faculty control, increased need for staff development, and the addition of specialized support staff.

4. Access to distance education may be a func-

tion of the wealth of students, rather than of institutional commitment, especially in regards to interactive technology such as computers and videodiscs.

Distance learning specialists also need to provide flexibility into the design of the courses. For example, adult learners have increased motivation when they have been given some freedom to select media and learning activities (Hull and DeSantis, 1979). This approach is compatible with the major theories in the field of instructional technology.

In conclusion, distance education is at the beginning of a challenging new era. We are moving away from the static, one-way instructional system so common at many institutions to new, exciting delivery media which will provide positive learning outcomes. Interactive systems will be the dominant distance learning delivery mode in the near future.

□

References

Bates, A. *Appropriate Teaching Functions for Television, Radio and Audio-Cassettes in Open University Courses.* Washington, D.C.: ERIC Clearinghouse on Higher Education, 1979. ED 227 805, 2-3.

Evans, A. Media Managers and Distance Education. *Media Management Journal*, 1986, *5*, 22-23.

Feasley, C. *Serving Learners at a Distance.* Washington D.C.: ASHE-ERIC Higher Education Research Report No. 5, 1983, p. 8.

Gray, R., and Martin, W. Starting a Microcomputer Staff Development Program. *School Learning Resources*, 1984, *4*, 7.

Gray, R., and Martin, W. Computerphobia, Human Relations, and the Media Specialist: In-servicing with a Full Hand. *International Journal of Instructional Media*, 1983, *11*, 280.

Gray, R. Videodisc Technology: Pros and Cons. *International Journal of Instructional Media*, 1978, *5*, 156-157.

Heinich, R. *et al. Instructional Media and the New Technologies of Instruction.* New York: John Wiley and Sons, 1986.

Hull, G., and DeSantis, V. How to Teach Adult Learners on Their Own Terms. *Audiovisual Instruction*, 1979, *24*, 15.

King, B. *et al. Support Systems in Distance Education.* Washington, D.C.: ERIC Clearinghouse on Higher Education, ED 227 747, p. 7.

Lewis, R. *Meeting Learners' Needs Through Telecommunications: A Directory and Guide to Programs.* Washington, D.C.: American Association for Higher Education, 1983.

McIntosh, N., and Woodley, A. *The Suitability of Non-Traditional Distance Learning Systems for Different Types of Students: The Experience of the Open University of the United Kingdom.* Washington, D.C.: ERIC Clearinghouse on Higher Education, 1980. ED 189 967, p. 5.

Seeff, A. Curriculum Issues in Telecommunications. *Current Issues in Higher Education.* Washington, D.C.: American Association for Higher Education, 1981.

Stevenson, W. *Britain's Open University, Use of Mass Media.* Washington, D.C.: ERIC Clearinghouse on Higher Education, 1981, ED 231 310, p. 11.

Turner, J. Computers Are Said to Be Little Used on Many of the World's Campuses. *Chronicle of Higher Education*, 1986, *32*, p. 2.

Walker, D. Tight Budgets and Socialistic Image Plague Britain's Open University. *Chronicle of Higher Education*, 1984, *28*, p. 33.

Building Connections for the Growth of Distance Education

Colin W. Dunnett

When discussing education within education departments generally, the words "distance education" have an immediate connotation relating to isolated students. Such persons are generally said to be disadvantaged due to their physical distance away from what could be taken as the norm in the provision of facilities for all aspects of an educational system.

What Is Distance Education?

There are other needs that can now be satisfied by "distance educators." Since the introduction of modern telecommunications, we have begun to realize that some very urban problems generated through declining enrollments as well as many other situations which put students at a disadvantage can be considered as able to be helped by distance education methods. These situations are not necessarily a result of the physical separation of pupils from normal facilities. Many such situations can be solved by the application of a program involving telecommunications. These solutions in total have become known as the process of Distance Education. That process of Distance Education is a total application of all of the education processes in circumstances where the receiver, i.e., the pupil, *is not face-to-face with the teacher.*

Perhaps we should coin a phrase other than "distance education" to imply non face-to-face teaching. Such a phrase would have to avoid the connotation of distance and yet include it without the introduction of some other equally misleading concept.

Alternatives could be "remote controlled education," "education by telecommunications," or perhaps "external education." However, I believe all of the alternatives to be limiting and equally capable of misinterpretation. It is now more easily understood, and I believe necessary for us to get

Colin W. Dunnett is Principal Educational Officer, Educational Technology Centre, South Australian Education Department, Adelaide, Australia.

used to the phrase "distance education" *implying all forms of education in which the teacher is involved and operating specifically in a role that is not in the classroom or in the same location as the pupil.* Such a phrase will include all of the situations involved in the problem.

Terms such as "open education" have been offered as suitable alternatives. This deserves a special mention, as it is the description of a style which will generate a situation within a traditional situation as opposed to describing a method of delivery or solution. Open education may well involve distance methods, but I feel each term constrains the other and confuses the argument or thesis if used in the same context as distance education.

The term must imply an educational method of solution, and not an event, if it is to be all-embracing as a solution of many problem areas or situations. Having hopefully clarified that definition, the actual activity requires a little more discussion before describing the process of laying foundations for a system that will enable those activities to expand and become the norm.

Forms of Distance Education

Perhaps the very first forms of distance education were those activities which the classroom teacher suggested should take place outside the classroom. When children were invited to do something specific at home or watch a particular program in the evening, visit somewhere, ask somebody something: then such activities could be described as processes involving the student in non classroom or non face-to-face action. This process then falls into the area of activity described above.

It is then only a small step to realize that the provision of resources for the classroom, and delivered to the classroom for use, are instruments of education within a distance education mode. Such resources are the tools of all kinds of education, both distance and remote. However, those tools are vital in that they add the dimensions of vision, movement, and sound to many interactive teaching modes which would otherwise be just a single component activity. The resources become absolutely linked to the process; although able to be used independently, they are created and designed primarily for a distance mode.

Perhaps the "ideal" mode of distance education is two-way television, with the pupils being able to see the teacher and the teacher being able to see the pupils in almost a face-to-face situation. It replicates what goes on in the classroom, precisely with perhaps only the immediate environmental effects being missing, and has all the "educational"

Telecommunications for Learning

17

resources of the teacher's choice and need available to the pupils.

Even those missing environmental and local effects could be provided if they were necessary but this is where trial, evaluation, and cost-effectiveness must begin to be considered together. While it is desirable to have a teacher in a face-to-face situation in every classroom for every subject for every pupil, such luxury can never totally exist and we must evaluate the very many alternatives that are now available. However, sometimes a "need" may demand satisfaction regardless of cost. Perhaps we may be going into a situation whereby even interactive television is necessary as a remedial tool for those that cannot achieve an objective by a less expensive method.

Building Connections: The Railway Principle

When supplying a solution to a problem, which up until now had no possibility of even being addressed, we must not criticize such a solution in terms of the non-achievement areas of the distance education method. Instead of criticisms of non-achievement, we must evaluate the components of the actual achievement in terms of their cost-effectiveness. Clearly, almost always, achieving something is of infinite value compared to a system which tackles nothing.

When accepting the principle of distance education and the use of technology generally, there inevitably occurs a problem in redundancy of equipment, i.e., as fast as one innovates and establishes any system using, say, a particular telecommunications device, then that telecommunication device itself can become outdated. The replacement of a device must occur and be part of a future system. It is very much like the videocassette recorder situation. I do not believe we will ever have a single standard of video recording in an Education system. Equipment standardization will fail because as fast as we attempt to replace old models, new equipment arrives. It is equally so in distance education, and any designed network or system must be built around a continuing changing scene of technology.

It is possible, however, to lay the groundwork of a network of connections on the railway line principle, making one's decisions in the early stages that the delivery will be on a specific gauge, in this case, a narrowband system of audio frequencies. Eventually, technology will provide new systems that can be carried on those railway lines. In building that railroad we simply are then able to wait for the design of the railway trains and railway engines to accommodate the communication which we wish to carry in any mode. We will, however, have begun a railway system plan of construction.

It is vital that the communication containers at the moment fit today's carriers. To continue to worry about alternative media is synonymous with the argument of the motor car versus the railway. The paradox of such an argument is illustrated when flat-top railway wagons are used to carry motor cars into central Australia, where the railway system, as a long distance carrier, is more cost-effective than the motor car. Such "cooperation" of systems is not only commerical but obvious.

This merely supports the need for both systems to have an area of overlap or compatibility. Hybrid systems of communication will be inevitable and any attempt to build an "all or nothing" technology of any kind will be doomed, not only because of cost, but because of the equipment redundancy which would be inbuilt into such a decision.

Let us go ahead and build these railroads which will bring to education a flexibility and adaptability that has been unknown in its history. Such communication systems will form the basis of a new society. Communication is at the heart of education, and communications technology will make the benefits of learning available to all who need them. □

Telecommunications and the Building of Knowledge Networks: Here Today, Much More Tomorrow

Dennis Adams and Mary Hamm

Electronic knowledge networks are an important part of the future for educational technology. The world is being served by a rapidly expanding grid of satellites and computer accessed networks. By tapping these possibilities schools can expand the educational horizons of students whose education is limited by schools too poor or too remote to do a good job of teaching certain subjects (from Physics to Art History).

Like other important technological applications, electronically accessed networks of knowledge will take some time to fully develop. The changes brought by the automobile, for example, took some forty or fifty years to unfold. Technological development time is now more compressed, but it still takes five to ten years for something to get off the ground. With the quantity of information we must deal with doubling every few years, computer-controlled networks are making an impact *today*. *Tomorrow* we will be able to simultaneously access text, images, and sound over the same line. By the 21st century electronic knowledge highways will merge at the *individual* and at the *world*.

About a quarter of the 10 million computers in American homes—and more than half of those in business—are plugged into the phone system so that they can connect to online databases, electronic mail, and computer teleconferencing systems.

Almost unnoticed, nearly a million Americans use either *CompuServe*, *The Source*, or *Dow Jones* electronic networks. Many are telecommunications amateurs who use their microcomputers and phone lines to access everything from international news, to movie reviews, to research on topics of special interest. Thus computer accessed databases are starting to quietly reframe the way individuals access world knowledge—*and* the way knowledge from the world has access to them.

Today's networks have problems. The sources of information are few. Many of the networking programs are too complex for an amateur to just pick up and use. Commands are often difficult to understand. Logging-on takes too much time. Modems are slow. And no one seems to be teaching computer users how to take advantage of telecommunications possibilities. Without electronic networking the computer is underutilized, if not isolated.

Like the automobile analogy used earlier, electronic networks need easy-to-use equipment and "good roads" to give us the capacity to transport the best teacher and learning activities from one site to hundreds (simultaneously). The technology is here, we just have to connect it. In a few years we will have more sources of information and equipment that is easier to use. By the 21st century knowledge networks will commonly challenge students to think and explore with thoughtfulness and imagination. The question is: How do we get to intelligent superhighway knowledge networks by the early part of the next century?

Exploring human machine sensory and cognitive systems—and the process of natural interaction—is the best way to figure out how things can be done. We are within ten years of breakthroughs in artificial intelligence that can help us create machines that are capable of exploring networks of knowledge on their own—and deciding what information to bring back to students. In just a few years we will have computer-controlled communications technology capable of tailoring information (text, imagery, and speech) for individual human users.

"Hypermedia"

"Hypertext" or "hypermedia" are terms that are just beginning to creep into our technological vocabulary. What does it take to make information "hyper"? First we need software that can help spontaneously gather bits and pieces of information in associative fashion. Secondly, we need computer storage devices (like videodiscs or CD-ROMs) that allow us to randomly access text, pictures, speech, or numbers. Such information can be transmitted anywhere—to the computer standing next to the optical disc player or to a PC halfway around the world.

Hypermedia programs allow us to create a window on a computer screen and selectively hop down any database alley that is on-line. Scholars have always used a card catalog and educated hunches to sort through information. This has

Dennis Adams is Associate Professor, Educational Technology, San Francisco State University, San Francisco, California. **Mary Hamm** is Associate Professor, Elementary Education, at San Francisco State.

not always been sequential or linear. Hyper-media greatly accelerates and amplifies this process.

"HyperCard" is Apple Computer's contribution to building hypertext. A related example is the "Microsoft Reference Library" that some colleges and high schools are using. This gives students a taste of what hypermedia will soon be like. Among other things, it gives them random access (using CD-ROMs) to the full text of *Bartlett's Familiar Quotations*, the *Chicago Manual of Style*, the *World Almanac*, and the *American Heritage Dictionary*.

Links to databases can put many libraries or information utilities within windowing distance of a personal computer. Information can be made available in seconds—wherever it resides. Soon we will be able to place our cursor on a footnote, click the mouse, and zoom in on a full text of the reference from a distant database. This kind of travel on knowledge highways requires students to act as much like *researchers* as traditional *learners*.

Hypermedia can also give users fresh, fast, free-associative images that are much like an action painting that you can feel as they are like printed text that you can intellectually comprehend. The overriding goal is to teach students to use all of their mental resources to think effectively. We have already seen the arrival of less complex PC accessed communication networks that have made those using them broader in their skills and interests.

Microcomputer Accessed Telecommunication Networks We Can Use Today

Electronic Mall (e-Mail) uses a computer much like a telephone, except that messages are typed on a keyboard and the print appears on a distant computer screen. Hooked into a telephone modem, computer messages can be accessed instantaneously—or "picked up" (viewed) at the receiver's convenience, like regular mail. Electronic mail erases the constraints of place and, to a large degree, those of time. It is estimated that in the United States over 500 million messages a year are sent via electronic mail.

Documents can be *shared* on any subject. People in different locations can collaborate in writing, share research, or jointly study (and comment on) a document. In a recent experimental program, inner-city children and children in rural classrooms drew storyboards of life on the prairie and on city streets. The text and illustrations were shared electronically.

Conferencing allows groups in diverse locations to take part in live two-way communication. Conferencing is particularly effective with the new computer-based video devices that allow participants to see one author on a split screen. Inexpensive digital compressors can be attached to a microcomputer, allowing color television pictures to be transmitted computer to computer, on regular telephone lines. Another approach is satellite teaching. The TIE-IN Network (operating from Texas) is a good example; lectures are beamed to a satellite and down to school sites around the country.

Library Databases contain both documents and software. Encyclopedias, for example, are available on disc or on telecommunication lines. Students can upload and download software or documents. Publishers are starting to explore these library databases as a distribution outlet.

Adding Artificial Intelligence to the Knowledge Grid

Some databases are constructing artificially intelligent expert systems that allow microcomputer users many miles away to interact with videodiscs with stored images of notable scientists (doing experiments) and artists (explaining various creations). The goal is to convey the way in which these human artists or scientists think and work. The result can be a total immersion in the creative situation under study. Artificially intelligent (computer-controlled) TV sets can also attend to learning style preferences by combining databases and the networks to assemble programs that reflect a viewer's interest.

Bell Laboratories has developed a number of microcomputer based teletraining application programs for its ALLIANCE telecommunications network. Selected schools around the country have been linked with interactive graphics, audio, and print via telephone lines. The AT&T public network allows students to interact with taped lectures and other instructional material. For example, students can ask questions and interact with a recorded image of a teacher or subject matter expert. When simulation is added to this mix, students can enter an electronic environment, experiment with a pioneer researcher (for example), develop a structure for asking the right questions, and graphically see the consequences of their decision making.

Rapidly converging electronic technologies are in the process of redefining communications media and learning. Technological developments in computing, television, interactive video, and telecommunications have the potential of dramatically transforming educational activities. At least one of the inexpensive possibilities—the electronic bulletin board—is already heavily used.

Electronic Networks Available Now

A microcomputer, telephone line, modem, and software program allow students and teachers to communicate with others in diverse locations through a vast and varied network of over 3,000 electronic bulletin boards and databases. The Word Processing Users Group (W/Pug) has set up the SCROLL Bulletin Board, which allows uploading and downloading of documents for evaluation and review. W/Pug also maintains a library of public domain disks which are available in more than 100 different computer formats. And it publishes a newsletter.

The Source, CompuServe, Specialist, TechCentral, DeafNet, and the *Dow-Jones News/Retrieval System* are examples of large-scale computer accessed telecommunication networks that also contain bulletin boards. These are like electronic magazines that can be used by schools, businesses, and government agencies. The Association for Educational Communications and Technology (AECT) is just one example of a professional organization that uses its *TechNet* computer-based electronic network to communicate with members and communications experts. Larger commercial databases, like *CompuServe,* have over 250,000 subscribers and 100-plus bulletin boards on just about any subject.

Anyone with a computer, modem, telephone, and a $50 software program can start an electronic messaging system. This gives microcomputer accessed bulletin boards the lowest entry cost of any mass communication medium.

Dailing the Future

We are heading toward individualized learning-research technology that provides access to the world of ideas based on our preferred learning style. If the schools are given the technological resources available today, teachers and students can take advantage of electronic tools for distance learning.

Easy access to faculty, other students, databases, and library resources will change the way that information is accessed and transmitted. Computer-controlled networks hold immense potential for a real revolution in learning through electronic collaboration. The ideal of life-long learning is coming within our grasp. The most up-to-date information and learning strategies can be accessed by alumni, schools, and industry. How will various electronic media and learning converge? How will the technology connect the human mind to global information resources?

We are now in the process of putting together the technological elements needed to give us computer access to a personalized set of learning experiences. We must spend at least as much time developing a modern philosophy of teaching and learning. Understanding the characteristics of effective instruction is an essential piece of the puzzle. The development of basic skills, habits of the mind, wisdom, and traits of character will be affected—one way or another—by the educational technology already on the horizon.

New media are being created that will transform human learning, expression, and communication. Much of the important information in the world is already in the process of being stored electronically. The challenge is to make sure that all of this information can be easily accessed (and cross-associated) *in* the right place *at* the right time. While learning to use what's available today, we need to start building the infrastructure of the personalized knowledge highways of the future. □

References

Adams, D., and Hamm, M. *Electronic Learning.* Springfield, IL: Charles Thomas Publisher, 1987.

Adams, D., and Fuchs, M. *Educational Computing: Issues, Trends and a Practical Guide.* Springfield, IL: Charles Thomas Publisher, 1986.

Adams, D. *Computers and Teacher Training.* New York: Haworth Press, 1985.

Gibson, W. *Count Zero.* New York: Ace Publishing, 1987.

Kobayashi, K. *Computers and Communications.* Cambridge: MIT Press, 1986.

Minsky, M. *The Society of Mind.* New York: Simon & Schuster, 1987.

Some of the concepts in the "Hypermedia" section were developed in cooperation with John Quinn, San Francisco State University.

Linking Teachers to the World of Technology

Gerald Marker and Lee Ehman

Helping teachers to use technologies has become part of our education agenda. In a recent survey, perceptions of education school faculty and student teachers showed that readiness to teach with computers is rated lowest of the 12 aspects asked about, with only 29% of the student teachers indicating they felt prepared in this area (AACTE, 1987). Twice as many education school faculty believed their student teachers were prepared to teach with computers. This gap in perceptions was the largest among the 12 aspects rated by both groups. A reasonable inference is that the same result would obtain for inservice teachers. We must work on increasing teacher preparation to use technology in teaching.

This article points out some teaching technology issues and lessons gleaned from our work in thirteen Indiana schools, and at Indiana University, during the past two years. It provides some ways of giving teachers the knowledge, skills, and attitudes necessary to use technologies effectively in their classrooms.

We first draw a picture of the "world" of public school teachers and administrators in Indiana, as they think about and use technology in their work. The article then builds upon this by describing a distance education network project—based on the AT&T electronic mail system—employed in Indiana and other school systems during the spring of 1988. We comment on some lessons learned about teacher knowledge, skills, and attitude. The article concludes with general implications from both projects for technology-related teacher education.

Status of Technology in Indiana Schools

During the spring of 1987 structured interviews were conducted in thirteen Indiana school districts and with officials of the Indiana State Department of Instruction. All school districts were members of the Indiana Public School-University Partnership,

Gerald Marker and Lee Ehman are on the faculty of the Indiana University School of Education, Bloomington, Indiana.

a part of the National Network for Educational Renewal, and ranged in type from a large urban school district to a small, rural district. Visits were typically one day in length and persons interviewed included superintendents, principals, computer coordinators, teachers, librarians, and media specialists. Over 70 interviews were conducted, and interview questions were organized under broad categories such as: hardware, software, professional roles, support services, curriculum, problems, trends, and dreams. The comments which follow are based upon data collected during this study.

Diversity of Technology Settings: While many teacher educators have been frustrated by the slow pace of educational change, they have also profited from it. Until recently, stability in the workplace has allowed easy prediction of the settings in which teachers would work. This stability has diminished; educational settings are changing and training programs must now change more quickly as well.

Diversity in the workplace is a fact of life for teacher trainers, making it difficult to predict the settings in which new graduates will work. For example, should secondary education programs be designed on the assumption that graduates will encounter an MS-DOS or Apple computer environment? What should trainers assume regarding teachers' access to departmental computers, computer labs for their students, and software? The general educational reform underway in this country calls for greater emphasis upon problem solving and analytic skills, but most school software collections rely heavily upon drill and practice materials. Teacher utility software such as gradebook programs and student-record data bases are provided in some districts, but are assumed to be teacher-purchase items in others.

Training programs cannot assume uniformity of personal support, either. The technology support teachers can expect varies from full-time computer coordinators and regular training sessions in some districts, to little more than general encouragement from building administrators in others.

Such wide variations in the technology settings of the workplace are a complicating factor, but trainers of teachers have no choice but to design programs that take such diversity into account. Even the technology-poor districts now expect new employees to possess basic competence in technology as well as a willingness to use what is available.

The Changing Role of Educational Technology: Industry and the military provide examples of where educational technology is headed. In those arenas, interactive videodiscs simulate training settings and two-way interactive video delivers instruction to remote locations. Such applications exist rarely in schools and would represent monumental changes were they to actually become commonplace. However, that should not mask the fact that widespread changes of a smaller magnitude are taking place in how technology is being employed in education.

Only a few years ago the emphasis in the schools was upon teaching students computer programming languages such as BASIC. Now programming has been replaced by teaching students to use specific integrated software packages, such as *AppleWorks* and the *PFS* series. Help screens and uniform command syntax and structures make the current generation of software less difficult to use for an increasing number of students, but even the least complicated software can seem intimidating to novices, especially if they are experienced teachers who feel pressured to learn quickly.

The nature of instructional software is also changing. While drill and practice courseware still predominates in schools, programs that promote problem solving are becoming available in good quantities. The use of instructional materials in videodisc format is still in its infancy, but burgeoning numbers of school VCRs enable wider use of television programming in teaching. Some educators hold high hopes for both technologies.

In many schools the vocational curriculum leads the way in applying technology. Many journalism students now use desktop publishing programs to prepare the school newspaper, and drafting students use CAD programs to prepare blueprints. Word processing and spreadsheets are now taught in many high school business departments.

Librarians are also feeling the impact of technology. Typically technology is applied first to library inventory and circulation control, but access to on-line and CD ROM data bases is beginning. Only a tiny number of libraries have data stored in a videodisc format.

A decade ago the focus of technology in the schools was on the hardware. Teaching students to program computers was assumed to remove the mystery of those machines. Today attention is on what the machines can do to help us with daily tasks, rather than on the inner workings of computers themselves. Increasingly, computer technology is taken for granted by teachers and students alike because it permeates the world around them.

Learning to Hit a Moving Target: The obsolete training program is an increasing problem for teacher educators as they struggle to keep pace with changes taking place in the public schools. The problem is compounded by the fact that we must predict what further changes are likely to occur in the workplace before new teachers move into jobs.

How do teacher educators cope with the "moving target" problem? First, they make assumptions. If MS DOS computers are predicted to become the standard in secondary schools, while Apple machines predominate in the lower grades, then training programs must be designed on the basis of that assumption.

Experience also supports the assumption that while new generations of videodisc players, CD-ROM units, and computers will find their ways into schools, they will not replace outdated equipment. Rather, they will take their place alongside the older models, further complicating the problem for trainers. The result will be the need for training situations which pay as much attention to the principles of technology (e.g., computer operating systems in general) as they do to training students to use a particular technology.

Predicting the technology component of the curriculum five or ten years into the future is only slightly less risky than making assumptions about what technology will then be in use. This particular area of the curriculum is changing so rapidly that the majority of school curriculum guides no longer depict current practice. Teachers who have experienced the revisions which are not reflected in the guides have the new course of study "in their minds," but the new graduate is likely to get little help from the outdated guides. Worse yet, they may have been prepared to teach a curriculum which no longer *exists*. New teachers who have had no experience with networking software or authoring languages are examples of the problem of the lack of fit between training programs and workplace demands.

Perhaps the most useful solution trainers can build into their programs is one dealing with attitudes toward future changes. Even if we are lucky in predicting changes in hardware and the curriculum, the fit between training and reality will not be perfect, and our graduates are sure to encounter the need to retrain several times during their careers. Training programs can anticipate this need by developing two attitudes in their graduates. The first is an expectation that the need to retrain is normal and that they can successfully apply the principles they already know to new situations. In short, they need to view the need to retrain as the need to sove a problem, and they need to have had experiences which make them confident that they can deal with such problems. The second attitude is the belief that results produced by applying technology to educational problems are worth the extra effort required. This attitude will result from learning about specific student learning benefits from employing technologies, rather than a set of untested beliefs that more technology automatically leads to better results.

Against this backdrop of the "world" of the Indiana school, we now turn to a specific example of how teachers in Indiana use telecommunications technology in their teaching, and examine the lessons learned from that experience.

The Long Distance Learning Network Project (LDLN)

What is the LDLN? The LDLN project was sponsored by AT&T during the 1987-88 school year to determine the effectiveness of long-distance learning in schools. Teachers, and the students in their classrooms, were linked electronically to teachers and students in other schools to form 37 "learning circles" that focussed on particular topics, such as careers, weather, and fiction writing. Information was exchanged among learning circle members via electronic mail messages. Teachers from grades four through twelve in over 100 schools in Australia, Canada, France, the Netherlands, West Germany, and the United States participated. Locations in the U.S. included California, Indiana, New York, New Jersey, and South Carolina. Within Indiana, there were 34 different schools in eleven school districts, which involved 61 teachers and about 1,500 students. Personnel from Indiana University also participated in support of the Indiana portion of the pilot test.

In Indiana, planning and teacher recruitment/training were carried out during August through December, 1987. Formation and startup of the learning circles occurred during January and February, 1988. The operational part of the project, therefore, happened between February and May, with only part of February actually available. Based on a largely positive external evaluation report, the project was continued by AT&T during the fall, 1988 school semester.

Lessons Learned in LDLN

Startup, support, timing, and implementation lessons: One obvious lesson we learned was not to begin use of a complex technology in an unrealistically short time period. We were too optimistic in assuming that teacher recruitment and training, lesson development, software development and support, and learning circles organization could be done from December to February, with all of the intervening events in the school calendar during that time period.

A second important imperative is to incorporate local support persons to lead teachers in their use of new technologies. Fortunately, we did plan for this resource by using coordinators in the eleven Indiana school districts, with university support of those coordinators. The result was a "safety net" of well-informed lead teachers who helped others.

The third lesson learned is to anticipate major "distractors" or "competitors" that impinge on teachers' time and effort. We didn't do so well in this regard. For example, a new state-wide competency testing program, ISTEP, was instituted in February during LDLN startup, forcing many teachers to devote most of their time and energy into preparing their students for test-taking, rather than beginning to implement LDLN objectives. No one could fault the teachers for this—it was an important priority throughout the state of Indiana. Another example had to do with vacations. Spring breaks came on different schedules, with the result that smooth information flow within learning circles was interrupted. We needed to anticipate these problems, but didn't.

A fourth lesson learned is when using complex technologies, provide high task structure for teachers. Translated into the LDLN experience, this meant that initial structure in learning circles— leadership, specific planning, and directions to teachers as to how to proceed—was critical to success. What we found out is that we can't expect teachers to learn how specific technologies (electronic mail, modems, learning circles) work, and invent their classroom applications at the same time, without wholesale confusion and frustration by many.

Fifth, teachers have to "believe" in the potential benefits of technology use, in addition to understanding them and being skillful in their use. Creating this belief must be part of the adoption strategy. According to the formal evaluation report, LDLN managers did this. Ways that teacher beliefs in LDLN efficacy were reinforced included news articles, school newsletters, testimonials by principals at meetings, parental support, and perhaps most importantly, student enthusiasm.

Sixth, the LDLN project illustrated that teaching technologies must solve real classroom problems and add more value than they cost in time, money, and effort. They cannot constitute "add-on" problems in their own right. The LDLN telecommunication helped to address several real educational issues. For example, the network connected students to others' diverse views and experiences, whereas the typical classroom lacks this "window on the world." Students were excited by the learning going in their circles, partly by the connection to students in other states and countries. Teachers' isolation, certainly one of the key difficulties in the profession, was alleviated through communication within the learning circles. Finally, the LDLN project was an excellent way to solve one of computer education's most vexing and real problems— what to do with the single computer in the back of the classroom!

Finally, educational technologies must be tools, not the objects of study, in the curriculum. One continuing issue in education is curriculum integration; the computer sometimes detracts rather than adds to this integration. The LDLN successfully integrated technology with the study of subject matter—science, foreign language, social studies, and English, through the learning circles concept.

Guidelines for Teacher Education

What can we glean from these two experiences to form useful suggestions for teacher educators trying to influence adoption and use of technology? We have drawn out four ideas which seem to hold true in both the Indiana schools study and the Long Distance Learning Network project.

Don't Oversell Technology to Teachers: Teachers are great hype detectors, rightfully suspicious and critical of new instructional ideas. Most have seen overblown fads and unfulfilled promises come and go, while they persevere in the classroom, helping their students learn and grow with whatever works.

Like other human beings, teachers sometimes find themselves convinced by new ideas which hold out hope for solving entrenched problems, such as student apathy or difficult-to-teach skills. Teacher educators have the responsibility to avoid raising expectations unrealistically high, because disappointment and knee-jerk rejection of new practices follows.

Some teachers in the LDLN project assumed at first that they and their students would be communicating interactively with their counterparts in their learning circles "in real time." This was not possible, nor even desirable, in the project, and we found it important to clear up this misconception quickly in order to avoid misunderstanding and disappointment.

Another feature of the LDLN work did lead to some teacher disillusionment because of higher than reasonable expectations. When first becoming part of their selected learning circles, teachers expected that leaders of the circles would quickly provide the ideas and structure required to foster enthusiasm and learning among the students in their classes. But the leaders were chosen more or less randomly, and without specific training in how to lead an electronic learning circle. Of course, many such leaders did very well, and those circles were successful. However, not surprisingly, some leaders were not able to provide the needed stimulation and resources to carry the day, and these circles floundered, and sometimes perished. The "promise" of the learning circles and their leadership, even though only implicit, was oversold, and some teachers and students dropped out of the project as a result. Their resulting attitudes toward using other technologies must have been made more negative by this experience.

Another example comes from the study of Indiana schools. Some teachers and administrators in the Indiana schools study commented that computer labs in their schools had not lived up to expectations. Short breaks between class periods did not permit time for lab set-up. Heavy demand often limited classes to a once-a-week schedule, with students forgetting much of what they had learned prior to the next session. Machines and software often did not operate as advertised. Many teachers became skeptical and assumed that it was prudent not to count on their period-a-week in the computer lab as a reliable instructional tool.

Allow Twice the Time You Think You Need: Two aspects of allowing ample time for adoption of technologies have to do with implementing software and hardware systems, and training teachers (and sometimes students) to use them. In the LDLN experience, one key timing issue was the development, testing, and delivery of a communications and electronic mail software package for the Apple II series computer. AT&T already had such a package for the IBM and compatible computers, but because schools use the Apples so widely, a software firm was employed to develop the needed package. The original delivery schedule was for December, 1987, just a couple of weeks after teacher training was scheduled, and two months before the full in-class startup of the project. The Apple software delivery date slipped, and a bug-free version wasn't available until mid-February. Understandably, this timing problem hurt the project, and forced most local coordinators and teachers to adopt whatever communications software was close at hand, and to rely on the more complex command-driven version of the electronic mail package. Better planning and allowance for more cushion in the software delivery schedule would have saved many headaches and would have contributed to more positive outcomes.

Another part of the timing issue relates directly to training. In both the LDLN project and the Indiana schools technology study, we found the need for spacing of training, with teaching of a reasonable set of ideas and skills followed by some "soak" time and opportunities for practice and mistakes; only then should additional material be taught. We made the mistake of trying to "teach it all" in a single one-day session. Instead of the teachers knowing how to word-process and edit simple files, convert them to ASCII files, upload and send them electronically via mail, and read and store responses, which was the intention, some knew only confusion and anxiety about making it all work when they returned to their schools.

Technology Should Help Teachers Solve Old Problems: Some of us try to invent new educational problems that teachers don't yet have, and then "solve" these problems with technology. While there is certainly a great need to create ways of fostering learning that are so far undreamed of, teachers need most urgently help with the old problems.

One good example of this was the momentary fixation on teaching programming languages that many of us tried to foist onto classroom teachers. While teachers needed tools to help students learn to read and write—and the computer seems to foster these skills—many teachers were sold the idea that the "problem" needing attention was the lack of student understanding and skill in computer programming. Therefore, for a few years, much teacher education effort was placed in "computer literacy," involving programming as one important part. Most teachers, however, saw no worthwhile student outcomes from this emphasis. Subsequently, the pendulum has swung away from programming, and corresponding study of the computer itself, and now heads toward using the computer as a tool in learning skills and knowledge. We invented a problem that didn't exist, while teachers (perhaps not so patiently) waited for computers to be applied to teaching and learning problems more salient to the students' needs.

Another example of a solution in search of a problem was the provision of a "talk to subject matter specialists" service which was available on the LDLN network. Teachers apparently felt quite competent in their subjects while at the same time making extensive use of a learning circles process consultant. Much effort was spent in setting up this resource, which was simply not perceived as useful by teachers, and therefore not used by them.

Train for Evolution, Not Revolution: The point here is that rather than trying to change the educational world of the teacher in fundamental ways with technology, we should be training for adaptability by teachers to foster incremental improvement of their work. From the evidence we have, there is no sign among the technologically innovative Indiana school districts that any revolution is on the horizon, *nor is one wanted.* What is —and what can be provided by good teacher educators—is to promote flexibility and adaptability in teachers.

One often-cited problem in school is: What do I do with that single computer sitting in the back of the classroom? The LDLN project answers this question with a powerful and elegant telecommunications network linking those lonely, single computers with many others in the backs of classrooms around the world. But a creative teacher, working with other teachers, might carry out the same basic activities as in the LDLN project—without the modems, telephones, and electronic mail system. Diskettes can be passed, or mailed, back and forth among "learning circle" members if the sophisticated network is not available. The learning outcomes are the same; a dialogue of questions and answers can be promoted, data exchanged, others' cultures examined, all with the computer being used to assist students in composing and exchanging ideas. This is only one adaptation of the LDLN idea possible, and is within reach of creative teachers.

Conclusion

Our experiences in these projects confirmed other experiences we have had involving educational innovations. In many respects helping today's teacher use technology is not different than helping teachers in the 1960s learn to use the "new" math or "Green version" biology. What is different is how quickly technology is changing, and that technology itself is contributing to the rate of change. On-line databases and bulletin boards, world-wide computer networks, and CD-ROM information services are both solutions and contributions to the problems faced by teachers as well as teacher educators. At a time when the "half life" of technology is counted in months, it is easy to forget the basic lessons we have already learned —but that would be a mistake. Setting realistic expectations, carrying out incremental training with an emphasis on the basics, and focusing on user-perceived problems are principles just as important as we teach 1980s technology as they were when we were preparing teachers to use overhead projectors or reel-to-reel recorders. If we have a single message, it is that technology should not distract us from what we already know about training teachers. □

Reference

American Association of Colleges for Teacher Education. *Teaching Teachers: Facts and Figures.* Washington, DC: AACTE, 1987.

Literacy in the Electronic Age

Peter H. Wagschal

Introduction

Almost a decade ago, a very brief article of mine appeared in *The Futurist*, suggesting—among other things—that reading and writing do not play as central a role in American society now as they once did. The article drew a quick and furious response from John Simon in *Esquire* Magazine, and then seemed to settle into well-deserved obscurity.

In the intervening years, we have seen the computer craze peak and decline, the publication of several books on the question of literacy in America, a genuine glut of critical reports on the quality of education, and a steady, quiet, unheralded continuation of the trends which I described in that brief article. Thus, the time seems auspicious to raise some of those literacy-related questions once again, lest we all go off on another Educational Revolution that ignores the realities of our daily lives.

I have no desire to engage in a statistical war, but, no matter whose data you look at over the past 20 years there is one reality which simply will not go away no matter how much we academicians and literary snobs dislike it: Americans spend more time watching television than anything else they do, except for sleeping. While our television watching habits have their ups and downs, we have been consistent in watching somewhere between four and six hours of television—every day, 365 days per year—since the early 1960s. At the same time, our reading habits—which have their own high's and low's depending on which study you choose to believe—have simply not kept pace with our video addiction.

Most authors—myself included—who take notice of this incredible dominance of television in American life are quick to point to the evils of such a video-dominated lifestyle. Neil Postman's *The Disappearance of Childhood*, for example, places the television set at the center of the decline of virtually everything worthwhile in Western Civilization since the invention of the printing press. In Postman's scheme of things, it was the printed word which brought humankind out of the Dark Ages, and it is the television set which now threatens to dismantle childhood, civility, critical thinking, and maturity—all in one quick flip of the on/off switch.

Much of what Postman—and other authors who are critical of the Video Age—has to say about the effects of television is undoubtedly true. The impacts of all forms of electronic communication are staggering, and it will certainly be at least a century before some clever sociologists and anthropologists can tell us what all this television-watching and computer-using has done to our society. There can be no doubt that a society whose citizens' most frequent waking activity is sitting in front of a television set must be changed by that single fact, especially when the technology did not even exist as little as 50 years ago.

But—at least for my present purposes—all the furor and debate over what television does to us is largely irrelevant. In the first place, we won't *know* the answers to such questions within my lifetime, and in the second place, we will continue to watch an enormous amount of television regardless of those effects. Whether we like it or not, and regardless of what we believe its effects to be, electronic communications media are here to stay and they constitute the major vehicle through which all of us find out whatever it is we think we know about the world. We cannot spend as much time as we do in front of the television set without succumbing to its influences, if only by default. During the four to six hours per day that the Tube mesmerizes us, it has become our reality and other sources of ideas, information, values, and wisdom have little chance to compete with such a dominant medium.

Hazards and Benefits

The hazards of television-watching are so commonly mentioned as to need little emphasis, but let me summarize the more obvious liabilities:

- Television-watching is a passive, hypnotic activity which demands nothing of the viewer except laziness.
- Since television is primarily a visual medium, it makes analysis, critical thinking, and other acts of rationality and intelligence virtually impossible.
- Programming on commercial television aims at the lowest common denominator and hence rarely portrays situations with any richness or complexity.
- Since television programming is determined purely by economic profits, all other

Peter H. Wagschal is Director of Advanced Graduate Studies, School of Education, National University, Vista, California.

goals of a communications medium are perverted toward financial ends.

- Television relies on sex, violence, and the spectacular to capture and hold its audience's attention.
- Television is such a "fast" medium that difficult subjects cannot get sufficient time to be explored in any depth.
- Television trivializes everything: it becomes impossible to distinguish crucial, real events from fiction or even commercials since everything is treated in the same casual manner.
- Television is socially isolating.
- Television encourages simple-minded thinking, stereotypes, and the predominant pattern of sexual, racial, and socio-economic bias.

We will return to some of these hazards, but for the moment let me assume that they are all—at least to some extent—real, and suggest that even while the medium is fraught with such difficulties, it provides the following kinds of benefits at the same time:

- Television enables people to see and experience things, both within and outside their culture, that they would otherwise know nothing about.
- Television has provided more peoole with more knowledge—superficial though it may be—about the world in which they live than has ever been the case through any other information medium in the past.
- Television has played a major role in providing common ground—for conversation and understanding—among the peoples of the globe.
- Television promotes social cohesion and unity through the universality of its presence—"Did you see Dallas last week?" is an ice-breaker all over the planet.
- In an age of specialization, television is the major contributor to the fund of general knowledge which people possess.
- While much of the programming on television is simple-minded, most of the people who watch it are not. People who have viewed television four to six hours a day for their entire lives become sophisticated viewers and are not easily taken in.
- While the mental skills required of television viewers are not the same as those required of readers, they are not trivial and may be appropriate to times which are as fast-paced and rapidly-changing as ours.
- Television has a capacity for arousing our emotions which makes it far more impact-

ful and memorable than the printed word.

- Television is the closest thing we have to a "democratic" information medium in which money and power cannot buy additional knowledge.

While not all readers will agree with everything on both of these lists—and while I'm sure there are hazards and benefits not included here—the summing up of television's value and impact is clearly not an easy task. As with any major technological innovation, there are costs and benefits which go well beyond our ability to compute a total.

The Written Word: Hazards and Benefits

Having inspected some of the costs and benefits of television as an information medium, it seems only fair to turn our attention briefly toward a similar consideration for the printed word—a task rarely on the agenda of those who are currently concerned about literacy. To the advocates of print, the following are some of its most significant advantages as an information medium:

- Print is logical, sequential, rational. It encourages the development of healthy, critical thinking skills.
- Print enables the reader to explore a subject, idea, or situation to as great a depth as he/she wishes.
- Print is an active medium, requiring the reader to draw on his/her imagination, logical skills, creativity, and life experiences in making his/her own judgments about what is being read.
- While some printed works are aimed at the "lowest common denominator," print provides specialized materials for audiences with different levels of sophistication on a wide variety of topics/issues.
- The printed word serves to enrich and uplift the reader's spoken vocabulary.

At the same time, however, print faces a variety of limitations as a communications medium, including the following:

- The printed word is socially isolating. A person reads or writes in solitude.
- Print is a hierarchical medium—a small number of people decide what is worth publishing for a larger number of readers.
- Print is an extremely slow form of communication.
- Reading and writing are fairly complex skills which take a long time and considerable work to master.
- The printed word can capture neither the richness of a visual image nor the complex-

ities and nuances of human drama and emotion.

The Role of Literacy

Having looked briefly at some of the hazards and benefits of print and television as information media, the question still remains, "Why is it so important to us for Americans to be literate?" Given the enormous amount of time that we spend in front of our television sets, it would seem reasonable to claim that our need for literacy has little or nothing to do with how we spend our leisure time and, hence, that it relates most closely to the world of work. Apparently, for us to be successful workers, it remains crucial for us to be skillful readers and writers.

Before looking more closely at this claim—i.e., that reading and writing play a major part in much of what the American workforce does on a day-to-day basis—it is worth a moment to make a list of other, practical functions that require some degree of literacy. In contemporary America, for example, if you can't read, then you can't:

- Understand road signs (although this is changing as visual signs replace printed ones).
- Understand labels on foods or other products.
- Follow instructions that come with a variety of durable goods—VCRs, refrigerators, cars, etc.
- Fill out a job application.

When it comes to the world of work, of course, the extent to which literacy is important depends entirely on the kind of work we happen to be engaged in. The "proliteracy" argument would have it that the more complex, high-status, intellectually demanding and (presumably) financially-rewarding jobs require the greatest degree of skill with the written word, and vice-versa. Doctors, lawyers, professors, and corporate executives, for example, are presumed to spend a substantial portion of their on-the-job time in reading or writing and, what is more, to be more likely to be successful at what they do if they are highly literate than if they can barely scrape by in the world of printed words.

In general, I would imagine (though I have not seen a time study of the previously-mentioned professions that would back me up) that this argument has some truth to it. Given the current structure of information storage and retrieval, reading and writing are probably still involved in a substantial piece of time for many high-status careers, and possibly even more so for the mid-level jobs that support them. But I hasten to provide the following three real-life stories as examples of what *might* be going on right beneath our noses without our even being aware of it:

A friend and former colleague of mine, whose name shall remain anonymous to protect him from the crime of not reading, managed to make it through four years as an undergraduate in a very prestigious university *without ever reading an assignment*. To be sure, he is literate, and was called upon to demonstrate that capacity in written tests, but he always found ways—e.g., lengthy conversations with his friends—to find out the content of the books assigned to him without ever having to read any of them.

Another friend of mine—equally anonymous and equally skilled in literacy—has been a Professor for almost two decades. He teaches upper-level Graduate courses, publishes frequently, stays well-informed of the latest developments in his field, and claims that he has neither read a book nor written a word for the past five years. He dictates everything he writes, and gathers his information from *short* written documents (e.g., newspapers, magazine articles, abstracts), from television and movies, and from conversations with friends and colleagues.

Finally, there is my acquaintance who is an M.D. in Family Practice. Again, he stays up-to-date (primarily through attending lectures and listening to audio cassettes on his way to and from the office) and carries out his day-to-day routine (including the constant entries in patients' files through dictation) without ever reading or writing a word.

I would not claim that these examples are "typical," nor would I argue that my friends could have found themselves in the professions they are in without having first obtained a high degree of skill at reading and writing. My friends are not *illiterate*, they are *aliterate*, and I have a sneaking suspicion that they belong to a growing sector of the American population. However unusual they may be, my non-reading acquaintances pose a serious set of questions for those who make so much of the role of literacy in the modern American world. If Doctors and Professors and Undergraduates can do their daily routine without reading and writing, why should we insist that they acquire the skills of literacy?

The Politics of Literacy

The answer to such a question lies more in the realm of politics—i.e., in the territory of power and how it is distributed in American Society—than it does in the worlds of work or Public Education. Even Jonathan Kozol seems to recognize this reality in his call-to-arms, ILLITERATE AMERICA, but he then proceeds to give us reasons and methods for avoiding what he must know is a political reality: In contemporary America, literacy has become our most significant tool for minimizing socioeconomic mobility.

The single place in American life where literacy makes the largest difference, on a daily basis, is in *school*, and after decades of studies which show the close relationship between success in school and parents' socioeconomic status, it should not be a surprise to us to note that Americans are more likely to be illiterate if they are poor or dark-skinned. Kozol and others who argue that a nation-wide literacy effort will tackle this problem seem to me to have the facts backwards: The sixty million adult Americans who are "functionally illiterate" do not tend to be poor and non-white *because* they cannot read and write. Instead, they are likely to be illiterate *because* they—and their parents—are politically and economically disenfranchised.

Selective access to literacy remains one of the American Public Schools' major contributions to the "tracking" of students into appropriate occupations. By the time children have reached the end of their elementary school careers, the schools have selectively distributed literacy among them: students from wealthy backgrounds acquire a high degree of skill, and students from less wealthy backgrounds acquire proportionately less adequate literacy skills. From there on, the rest of the "tracking" process is fairly automatic, for as one's literacy skills are, so is one's progress through the maze of American schooling and, thenceforward, one's likelihood of acquiring a job with high status and large financial rewards.

The point is that, in a society which allows only a limited amount of social mobility, literacy is a powerful tool in distributing socio-economic power *even though the daily requirements of high status jobs no longer—in reality—depend heavily on reading and writing.*

The Mythology of Literacy

Hand-in-hand with the politics of literacy goes a set of conceptions about what it means to be a "literate" person which rarely are subjected to careful scrutiny. These beliefs about literacy have a long history that has evolved along with the development of print as the central tool for distributing information. Unfortunately, they have always contained a substantial amount of elitism and mis-truth. Even more unfortunately, these mythological attributes of the literate person have less and less relevance in a world where print plays a smaller and smaller role in disseminating knowledge among the world's population. While it certainly once was the case that the only way a person could know anything about a foreign cul-

ture, for example, was to either go there, talk to someone who had gone there, or read about it, such is no longer the case. But the popular conception of what it means to be able to read and write has not changed, and so we find ourselves continuing to apply labels to illiterate persons which simply are not accurate.

According to the popular mythology, a person who is illiterate:

- Is stupid.
- Cannot reason, think, handle abstractions.
- Doesn't know much about anything.
- Is incapable of making intelligent decisions as a citizen in a Democratic Society.
- Has questionable morals.
- Cannot be expected to show wisdom on any important topic.
- etc., etc., etc.

These biases regarding people who cannot read and write have, of course, never been accurate. Even during its heyday on the Planet Earth, print never had such a monopoly on knowledge, wisdom, and morality as to deprive so many billions of people from their full humanity simply because they could not read or write. In fact, there have always been more people on the planet who are *illiterate* than people who are *literate*, and it seems a bit flippant to "write off" the majority of the world's population so casually. We are willing to attribute a little bit of wisdom and moral character to such historical figures as Socrates and Jesus, neither of whom ever wrote a word, so perhaps we should make room for the slim possibility that—among those great unwashed masses of illiterate Americans in 1987—there might well be a few who know something and can think or reason or come up with an interesting idea or even engage us in a stimulating conversation.

My point here is not that literacy has no value in American life, but only that we have a set of biases regarding what it means to be literate which does not match the truth. I have met too many stupid people who are literate, and far too many illiterate people who are brilliant, to rest easy with the mythology. And when I read in Jonathan Kozol's book about the 60 million Americans who are "substantially excluded from the democratic process and the economic commerce of a print society," my mind leaps immediately toward changing that society so that it can make better use of the human energy which it is currently wasting.

Toward a New Conception of Literacy

In the world of the late Twentieth Century, we find ourselves enmeshed in a set of technological revolutions which are changing forever the manner in which human beings communicate with one another. Radio, television, computers, videocassettes, videodiscs, and interactive television have already taken over a substantial portion of the communication which used to be possible only through the printed word. While it is presently true that a major portion of the computer-mediated talking that goes on in the world is still based on the printed word, the rapid improvement in both voice-generating and voice-recognizing machinery that is currently taking place could easily make print totally obsolete by the end of the Century.

Unfortunately, our conceptions of what it means to be "literate" have not had a chance to keep pace with these technological developments, and so we still talk and act as if all knowledge worth possessing resided in books. As we make our way through this difficult transition, I would suggest that it is crucial for us to highlight the need for new kinds of literacy that are more appropriate to the communications media that we use now and will have to rely on in the future. Just as the printing press made the need for rote memorization relatively insignificant, but created a previously-unparalleled need for reading and writing skills, so do our current media diminish the need for reading skills and create a demand for new kinds of "literacy."

What I am arguing, first of all, is that *wherever print can be replaced by other media, it should be.* Thus, I find it hardly compelling to argue that everyone needs to read so that they can fill out job applications (especially when the jobs being applied for require no reading or writing!) when job applications *could* be done via voice-recognizing computers. And I doubt that it makes sense to invest the kind of time and money that we presently do in teaching children to read if the major object of that training is to enable them to read the ingredients in a can of Campbell's soup. Instead, let's insist that Campbell's and the supermarkets use their optical scan equipment to tell us what is in a can of soup, so that we can spend our time in schools working on some crucial and currently untaught literacies. For example:

- Who teaches our children how to distinguish reality from fantasy in the television shows they watch daily?
- Who teaches our children the skill of comparing television versions of police procedure with the real world?
- Who teaches our children the more general skills of "critical television watching?"
- As all of us spend more time interacting with computers, who teaches us the skills of "critical computer interaction?"

As we move faster and faster into the electronic age, the skills required to be informed, critical, responsible citizens of a democratic society become further and further removed from literacy as we have known it in the past. We may spend our time bemoaning the passage of the Age of Print, and clinging to a set of outmoded skills which we continue to insist on using as a way of distributing power and wealth in America, or we can move boldly forward. That would require us to identify a whole new range of information-gathering skills which apply to our new technologies, and make sure that they are fairly and equitably distributed to our children, while at the same time making sure that the artificial hold which print continues to have over power and status is allowed to wither away.

Conclusion: A Fable for Our Times

Once Upon a Time, there were two children born to families of very different position. The Rich Girl was taught carefully and patiently to read, though she found it boring and much preferred her television set. The Poor Girl was taught sloppily and impatiently to read, and so felt more comfortable in front of her television set.

When she grew up, the Rich Girl became a Rich Woman. She lived in a world of sparkling jewels, watched television a lot, and never had any reason to believe that there were people in the world less well-off than she. When she *did* read, it tended to be Romance Novels, all of which made little mention of The Poor.

As one would suspect, the Poor Girl grew up to be a Poor Woman. She lived in a world of grinding poverty, watched television a lot, and was constantly reminded of the difference between television fantasies about Rich Women and the Real World in which she lived.

One day, the Rich Woman's chauffeur made a wrong turn, and she found herself parked in front of the Poor Girl's supermarket. The Rich Woman rolled down her window to get a look at where she was, and saw the Poor Woman through her open window watching a television program in her living room. Recognizing the television program instantly as one she watched every week, she rolled up the window, turned to her driver, and said,

"Look at that, Ralph, she's watching 'Dallas.' Isn't it comforting to know that no matter where you go in this world, we all live pretty much the same lives?"

Meanwhile, the Poor Woman noticed the limosine parked outside her apartment and wondered why the television people who were in it had come to her neighborhood. □

Part II

A Variety of Applications of Telecommunications for Learning

Communications Satellites: A Rural Response to the Tyranny of Distance

Gregory Jordahl

Introduction

Today's communications satellites epitomize the emerging information age. Orbiting 22,300 miles above the equator, these versatile messengers transmit voice, video, and data signals over great distances, at high speeds, and in large volumes.

Rural areas have recently begun to scrutinize the instructional applications of this technology so suited to reaching multiple and geographically-dispersed sites. The purpose of this article is to provide an overview of several current satellite-based instructional systems and to assess their potential role in rural education.

Why Satellites?

A number of factors have recently converged to create the current high level of interest in satellite-delivered instruction. To begin with, there is the increasingly intractable combination of demands upon and deficits within rural education systems. Hobbs (1985) has described the problem in the following way:

> There is the matter of increased public concern with education generally, highlighted by publication of *A Nation at Risk*; most colleges and universities are increasing their graduation requirements; new job skills are being demanded in the transition to an "information age" society; the oversupply of teachers of a few years back is being replaced by an expected teachers shortage in many subject areas; and economic and agricultural changes are shrinking the tax base of many rural localities. Improved educational effectiveness has been added to maintaining efficiency as a component of the rural school problem (p. 2).

The two most common attempts to minimize the chronic constraints on rural education—consolidation and resource-sharing—often entail social and economic trade-offs that hamper their potential long-range effectiveness. As a result, more rural educators are turning in increasing numbers to technological options for distributing increasingly scarce instructional resources to remote and isolated sites.

Events within the telecommunications industry during the past decade have also contributed to the burgeoning use of satellites in rural education. Powerful communications "birds" have been developed and launched, allowing reception equipment to become less sophisticated and hence much less costly. A surplus of satellite transponder facilities and increased competition from fiber optic technology have combined to bring the cost of satellite communications within reach of most school districts. In addition, the deregulation of satellite communications has enabled service suppliers to more freely experiment with innovative applications and price structures (Dordick, 1986).

Educators have thus begun to see that satellites, with their ability to broadcast voice, video, and data anywhere within a broad geographical area, might be an ideal way to beam cost-effective instructional and staff development programming from one central site to literally any school in the country, regardless of its size or location. Early experiments with this concept were encouraging. One of the first instructional applications of satellite technology was demonstrated in 1972 when the University of Hawaii used NASA's ATS-1 satellite to network three of its island campuses for an audio-assisted library science course (Bystrom, 1987). NASA also participated in a variety of direct broadcast satellite experiments that utilized ATS-6 to provide instructional services to small towns in the Rocky Mountains, Appalachia, and Alaska during the mid-1970s (Levinson, 1985). Alaska later pioneered the first state-wide educational satellite network when the LearnAlaska system was established in the early 1980s (Holt, 1985).

The success of these initial efforts provided much of the impetus for a pair of educational networks that were inaugurated in 1985.

The Oklahoma Arts and Sciences Teleconferencing Service

In early 1985, a partnership was established that linked the resources of the College of Arts and Sciences at Oklahoma State University, the Oklahoma State Department of Education, and the Oklahoma public schools for the purpose of enhancing educational opportunities for young people in the region. Called the Arts and Sciences Teleconferencing Service (ASTS), the project successfully piloted a German-language course in the spring of 1985, using a blend of live satellite video, audio teleconferencing, and computer-assisted instruction.

Gregory Jordahl is a consultant based in Eugene, Oregon.

Among the findings from the pilot, conducted at a small school in the Oklahoma panhandle, were data showing student performance levels comparable to those of average college freshman taking an equivalent German course at Oklahoma State (Holt, 1985).

Buoyed by the success of its initial effort, the project began providing German instruction via satellite to approximately 50 subscribing schools in Oklahoma, New Mexico, Texas, Kansas, and Colorado in the fall of 1985 (Garrett, 1986). Subsequent years have seen annual enrollment increases, from 107 schools in nine states during 1986-87 to 170 schools in 14 states during 1987-88 (Walters, 1988). The project's curriculum has also expanded and now includes German I and II, advanced placement calculus, and advanced placement physics. Additional courses are currently being developed.

Each course is designed and taught by professors from the faculty of Oklahoma State in Stillwater. Live instruction is provided two to three times weekly via Westar IV to subscribing districts in the ASTS network; class supervisors—typically teachers certified in subjects other than those offered via ASTS—monitor the students in each outlying classroom. Each video presentation utilizes an audio bridge to a designated host school that allows students there to interact with the instructor while the other schools listen in. Additional opportunities for interaction are provided by a toll-free telephone line staffed by graduate students at Oklahoma State and by an electronic mail link to each professor (Arts and Sciences Teleconferencing Service, 1986).

Also integrated into the ASTS instructional plan is a computer-assisted instruction component that supplements the video lessons during the non-broadcast days of each school week. The German courses, for example, use either Apple II or TRS-80 computers to provide vocabulary practice and reinforcement by means of voice recognition software designed specifically for the course by the ASTS German professor (Wohlert, 1986). Similarly, instructional computer games provide enrichment activities for the physics course.

Costs for ASTS services are assessed on a per-course basis. During the 1987-88 school year, Oklahoma schools were charged $1750 per thirty-week course; schools in other states were charged $2000 (Oklahoma State University, 1988a). The fee gives subscribing districts the right to receive and make off-the-air copies of the video broadcasts, to receive and duplicate all written course materials produced by ASTS, and to obtain the computer software designed for each course. Each participating district is responsible for purchasing and installing its own satellite reception equipment, classroom

video hardware, and microcomputers, all of which typically costs between $5000 and $10,000 per school (Barker, 1987).

The TI-IN Network

The Texas Interactive Instructional Network (TI-IN) is a privately-owned, for-profit project that utilizes uplink facilities at the Region 20 Education Service Center in San Antonio and at California's Chico State University to provide four channels of daily live broadcasts via a GTE Spacenet II communications satellite. TI-IN went on the air at the beginning of the 1985-86 school year, broadcasting a range of instructional programming to approximately 50 schools in Texas as well as several districts in California and Arkansas (Babic, 1987).

In the three years since its inception, TI-IN's enrollment has increased to approximately 450 schools in 25 states (De Freitas, 1988). The project's curriculum has likewise been expanded to include over 20 high school credit courses in foreign languages, math, science, social science, and art appreciation. Along with these full-year courses, TI-IN also offers numerous supplementary programs designed for specific K-12 age groups, plus 400 hours of staff development programming and teleconferences.

The methodology employed in TI-IN's delivery system simulates the traditional teacher-led presentation and discussion model of instruction. Supervised by adult classroom assistants, students in outlying schools use cordless telephones to respond to their instructor, who directs each daily 55-minute lesson from a Region 20 studio in San Antonio. An additional interactive tool for some classes is provided by electronic writing tablets that allow written work to be transmitted throughout the network. Dot-matrix printers at each receive site are used to distribute class handouts, quizzes, and other supplementary material, thus providing a third link between student and teacher. Microcomputers, however, are not an integral part of TI-IN's instructional delivery system.

Unlike Oklahoma's ASTS project, TI-IN supplies satellite reception equipment and classroom video hardware as part of each subscriber's start-up costs, which average approximately $19,500 (Babic, 1988). A subsequent annual fee of $5,250 includes a warranty and maintenance agreement, technical upgrades, video tape backup service, and student enrichment broadcasts (TI-IN Network, 1987).

Other Emerging Satellite-Based Instructional Systems

The initial success of the ASTS and TI-IN projects has led to a series of similiar programs throughout the country. Washington, Missouri, and Kentucky have all initiated satellite-based learning projects, which are in varying phases of implementation.

Washington's STEP System

Headquartered at Educational Service District 101 in Spokane, Washington, the Satellite Telecommunications Educational Programming (STEP) project was introduced during the 1986-87 school year. STEP, which utilizes uplink facilities at Eastern Washington University in nearby Cheney, served 15 rural schools in the Spokane area during its first year. During 1987-88, approximately 50 subscribing districts in Washington, Oregon, Idaho, and Montana aired some combination of high school credit courses and staff development programming from the STEP schedule (Roscher, 1988a).

STEP utilizes a blend of live, daily one-way video and two-way audio via toll-free telephone to provide a curriculum that, during the 1987-88 school year, included advanced senior English, Japanese, Spanish I, Spanish II, and pre-calculus. First-year costs are set at $8,500, half of which includes reception hardware and installation. Subsequent annual costs will be based upon the total number of subscribers and the level of participation of each district; STEP officials estimate that these fees will not exceed $4,000 in most cases (Roscher, 1988b).

Missouri's Education Satellite Network

The Education Satellite Network (ESN), a project sponsored by the Missouri School Boards Association, was launched in March of 1987 (Thomas, 1988). Unlike the ASTS, TI-IN, and STEP systems, Missouri's ESN is not a producer of original instructional broadcasts, but instead serves primarily as a broker of satellite-delivered services (Gardner, 1988). From an uplink at the University of Missouri at Rolla, participating districts are provided access to a range of satellite programming from sources such as ASTS, STEP, the National Diffusion Network, C-SPAN, Univision, and the International Television Network.

ESN annual subscription fees—from $3,500 to $7,250—are based upon the amount and type of programming an individual school wishes to receive (Education Satellite Network, 1987). These fees include the installation and maintenance of specially-designed integrated satellite receive hardware, which the network provides on a lease-like basis to subscribing districts.

During its first year, ESN served 23 districts throughout Missouri. A $319,000 matching grant from the National Telecommunications Information Administration awarded in the fall of 1987 is expected to enable the network to expand by an additional 80 to 100 sites during the 1988-89 school year (Missouri School Boards Association, 1987).

Kentucky Educational Television

By far the most ambitious of recent instructional satellite plans is currently being implemented by Kentucky Educational Television (KET). A major goal of KET's ongoing $11.5 million project is to install satellite reception equipment at each elementary and secondary school in Kentucky by this year. Additional dishes scheduled for installation at the state's colleges, universities, and libraries will bring the total number of downlink sites to between 1500 and 2000 (Welch, 1988).

KET plans to produce most of the interactive programming for distribution on the statewide system in a new communications center to be constructed near the network's headquarters in Lexington. In addition, funding is currently being sought for a mobile uplink truck that will allow various special events to be broadcast live to receive sites. The project hopes to begin piloting several demonstration courses by the spring of this year.

The Star Schools Bill

The rate of satellite technology deployment in rural areas of the country is likely to be accelerated by the passage of the Federal Star Schools Program Assistance Act in December of 1987. With an anticipated five-year budget of $100 million, the Star Schools program is designed to promote the use of telecommunication technology in providing instruction in foreign language, math, and science to traditionally underserved populations (U.S. Congress, 1987).

Those eligible for funding under this legislation include public agencies with telecommunications network experience or a partnership comprised of organizations such as state education agencies, higher education institutions, teacher training centers, public broadcasting entities, and local school districts. These partnerships must be organized on a statewide or multistate level.

The Star Schools program requires matching contributions of 25% and stipulates that 50% of the funding be used to benefit economically disadvantaged schools. A total of $19 million was allocated for FY 1988 (Oklahoma State University, 1988b).

Administrative Issues Relating to Satellite-Based Instruction

The trends described above point to a high level of interest in satellite-based distance learning alternatives as a solution to the chronic constraints that continue to plague rural educational programs. Funding commitments being made at both the state and federal level represent a significantly high degree of legislative support for telecommunications as a teaching tool. The programs that have been implemented thus far rely on a carefully-integrated mix of technologies that together represent one of the most potent educational innovations in recent years.

The above factors appear to indicate that satellite technology could make an important contribution to education in rural and remote areas. In addition to increased educational opportunities normally out of the reach of rural districts, satellite supporters point to other benefits as well. Cost-effectiveness, for example, is an advantage that is very attractive to most budget-constrained administrators. In addition to offering small-enrollment, advanced courses at a fraction of the cost of a teacher's annual salary, rural school districts also are able to save significantly on travel expenses associated with staff development by subscribing to the inservice programming that many of the satellite networks provide.

Holt (1985) has listed other major benefits. These include professionally-developed courses guided by master teachers with expertise in a particular area; the availability of tutorial assistance through electronic means; student access to high-quality print and electronic media materials as well as to a variety of information resources and data bases; and the integration of professional testing and evaluation with instructional activities, which guarantees a minimum statewide standard of performance.

However, in spite of these advantages, several issues will need to be addressed before the continued viability of satellite-based instructional systems can be accurately assessed. To begin with, there are as yet no published reports that measure the quality of the courses currently taught with interactive satellite technology. Though such data is forthcoming, its interpretation will be problematic since many participating schools reserve satellite courses for honor students who would likely show achievement gains regardless of the instructional method employed. Since other combinations of instructional technologies could provide similar, yet less-expensive interactive capabilities (videotape plus telephone, for example), the burden will ultimately be upon network providers to demonstrate that live video yields significant instructional benefits since that is where the bulk of initial and ongoing costs lie.

Another factor that will become important after the appeal associated with "space-age" tools begins to wane is student motivation. Traditionally the weak link in most distance learning systems, motivation is especially crucial in the types of advanced placement courses currently supplied by today's educational satellite networks. Administrators will need to carefully choose a classroom supervisor who has the talent and tact required to help students adapt to a class that relies on an initially intimidating array of technology and that requires a high level of individual initiative and personal responsibility to complete. The success of satellite-mediated instruction will hinge in large part on the quality of this human interface, particularly in those schools that open satellite courses to students at all ability levels.

Ultimately, however, the factor that will determine how well satellite systems are integrated into rural school districts is the level of support and direction supplied by administrators at each outlying site. Even though satellite technology is currently being embraced at a rapidly-expanding rate, its long-range success will be determined by how well administrators manage the fundamental change that the satellite dish represents for rural education. Human elements such as staff resistance and student apathy will limit the success of even the most well-designed distance learning project without strong administrative leadership.

Thus, those satellite-based distance learning projects that are most successful will pay close attention to the administrative strategies required at subscribing schools, particularly when the inherently-motivating newness associated with satellite delivery is no longer a factor fueling student enthusiasm. With careful management of these powerful new instructional opportunities, rural school districts may find themselves in a position to make significant and sustained improvements in the quality of their curricula—certainly a compelling alternative in an era of educational limits across much of rural America. □

References

Arts and Sciences Teleconferencing Service. German by Satellite. Information sheet included in subscriber packet published by Oklahoma State University, 1987.

Babic, J. Operations Manager, TI-IN Network, Webster, Texas, telephone interview, April 27, 1987.

Babic, J. Telephone interview, March 7, 1988.

Barker, B. Interactive Learning by Satellite, *The Clearing House*, September 1987, 13-16.

Bystrom, J. Founder of Pan-Pacific Educational and Cultural Experiments by Satellite Project, Honolulu, Telephone interview, May 17, 1987.

De Freitas, C. Educational Consultant, TI-IN Network, Mission Viejo, CA, telephone interview, March 19, 1988.

Dordick, H. *Understanding Modern Telecommunications*. New York: McGraw-Hill, 1986.

Education Satellite Network. Costs brochure included in subscriber packet published by Missouri School Boards Association, 1987.

Garrett, S. Director of Rural Education, Oklahoma State Department of Education, Telephone interview, June 5, 1986.

Gardner, H. Director of Satellite Communications, Education Satellite Network, Missouri School Boards Association, Telephone interview, March 9, 1988.

Gudat, S. Satellite Network Helps Keep Rural Schools Open. *Phi Delta Kappan*, March 1988, 533-534.

Hobbs, D. Bridging, Linking, Networking the Gap: Use of Instructional Technology in Small Rural Schools. Paper presented at the Department of Education's National Rural Education Forum, Kansas City, August 1985.

Holt, S. Wiring Rural Schools into Educational Reform. *Education Week*, August 28, 1985, p. 36.

Levinson, C. Education by Telecommunications at the Elementary and Secondary Level: Practices and Problems. *T.H.E. Journal*, April 1985, 71-73.

Missouri School Boards Association. (Satellite Network Receives Grant, *Board*, 1987, p. 1.

Oklahoma State University. Programs, *Learning by Satellite Newslink*, February 1988a.

Oklahoma State University. Star School Bill Passes Congress. *Learning by Satellite Newslink*, February 1988b.

Roscher, T. Director, Satellite Telecommunications Educational Programming Project, Educational Service District 101, Spokane, WA, telephone interview, March 21, 1988a.

Rocher, T. Personal correspondence, April 6, 1988b.

TI-IN Network. Costs for Network Subscription, mimeographed fee schedule, 1987.

Thomas, N. Education Satellite Network Is Launched in Missouri, *Learning by Satellite Newslink*, February 1988.

U.S. Congress. Title IX—Star Schools Program amendment to the Education for Economic Security Act, 1987.

Walters, L. ASTN German Course Coordinator, Oklahoma State University. Telephone interview, March 3, 1988.

Welch, S. Deputy Director, Kentucky Educational Television. Telephone interview, March 7, 1988.

Wohlert, H. German by Satellite: General Guidelines for Language Coordinators and Administrators. Oklahoma State University. Mimeographed syllabus, 1986.

Telecommunications in the Classroom: Can It Be Done? Should It Be Done? An Essay on Possibilities and Frustrations

David W. Swift

Modern technology offers marvelous possibilities for educators, but various obstacles may prevent us from using it. Many schools and universities do not provide the necessary equipment or technical assistance, thereby leaving it to individual teachers to acquire these on their own. Unfortunately, most teachers lack either the money or the expertise necessary to obtain and use the technology by themselves.

What, then, should a concerned instructor do? The easiest course is to do nothing. If, nevertheless, he or she persists in trying to get access to new technology, what obstacles will be encountered? Can he or she succeed? What price will have to be paid? Will it be worth it?

To answer such questions, or to begin to answer them, I will examine my own experiences in linking my class at the University of Hawaii with another, at California State University at Northridge, near Los Angeles.

I had in mind two-way communication, the opportunity to interact with a master scholar, to ask him questions about points I did not understand, or about facets of the subject which he had not discussed. Of course, for decades we have had movies and radio, and more recently, television and videotapes. Satellite relays enable millions to watch such programs as "Sesame Street" or Carl Sagan's "Cosmos."

But, wonderful as these may be, they do not provide for *interaction*. We cannot talk back, interrupt, and ask questions. It is a one-way process, prepared ahead of time, by persons who do not know us and who cannot foresee all the questions which might arise as we listen to or

David W. Swift is Professor, Department of Sociology, University of Hawaii at Manoa, Honolulu, Hawaii.

watch their presentations. Instead, we play a passive role. Our only choice is to watch or not to watch, or to listen or not to listen.

How does this concern us?

The Problem

The geographical isolation of Hawaii makes it difficult for students and teachers, as well as off-campus technicians and professionals, to keep up with the latest developments. This difficulty is most obvious in rapidly changing scientific and high-technology fields, but it can also be a problem in many other disciplines, including the arts, humanities, and social sciences. Aside from a few happy exceptions in which our location has aided us, scholars at the University of Hawaii face a major disadvantage in competing with those on the mainland of the United States.

Modern telecommunications technology could do much to overcome this disadvantage, but the University does not provide adequate facilities. Both its PLATO computer system and PEACESAT radio could be useful, but in actual practice, the installations are not. A third possibility, the telephone, also could help, but the problem with it is the high cost of long-distance calls as they are now handled by the University.

Aggravating the situation caused by this lack of facilities is the absence of a telecommunications advisor—a person whose skills and experience are broad enough to provide practical assistance for projects like mine. Although there are a number of specialists on the campus and in the community, particularly in computer conferencing, I did not find any one person who could advise me on the details of combining two or more modes, such as voice and visual, telephone and computer. Consequently, I had to feel my way through the tangle of technology, economics, and bureaucracy.

The lack of funds, facilities, and expert advice were serious obstacles, but nevertheless I proceeded to look for low-cost (or free) links with the mainland. I felt that such communication was important, both for myself and my students, and I wanted to demonstrate to the university community that it could and should be done.

My general objective was to connect students and teachers here with an expert at a world renowned center of activity. For art classes, this might be New York. Students of politics and government might focus on Washington. For aerospace, Los Angeles is a center of activity.

Planning the Project

The possibility of aerospace in Los Angeles seemed particularly well suited to this project. I had long been interested in aviation and space.

Southern California has the advantage of being the part of North America closest to Hawaii, thus presenting lower communication costs than an East Coast destination would require. NASA's involvement in space flight suggested that I could get their assistance, as well as that of Jet Propulsion Laboratory in Pasadena, the center of operations for much space research, including the Voyager flights to Jupiter and Saturn. Perhaps most important of all, one of the leading researchers in the fledgling field of aerospace sociology, B.J. Bluth, teaches at California State University, Northridge, near Pasadena. During the past two summers, she has taught a course titled "Update on Space." Among the many eminent guest speakers who have addressed her classes are Dr. Krafft Ehricke, President of Space Global, Inc.; Dr. Richard Johnson, Chief of Biosystems Division, NASA Ames Research Center; George Butler, Director, Advanced Space Concepts, McDonnell Douglas Astronautics; Charles Gould, Project Manager, Advanced Programs, Rockwell International; and Captain Stan Rosen, USAF, Chief of Spacecraft Systems, Defense Satellite Communications.

I had wanted to attend this course in order to expand my own knowledge of current developments in aerospace research, so when I learned that Dr. Bluth would be offering a seminar on the Sociology of Groups in Space, I asked her if I could listen in, providing that I could arrange the telecommunications. She agreed.

Then I thought that, since we were going to this trouble, why not invite other interested people in Hawaii to listen also? This subject matter has an appeal extending beyond the boundaries of any one discipline. Therefore, I thought that participants might include undergraduates, graduate students, and faculty in such fields as astronomy, aerospace studies, and engineering as well as the social sciences. Professionals and technicians from the armed forces and business might also be interested.

I offered the course as Sociology 495E, part of an open category established especially for innovative classes. I knew that unforeseen problems would arise, so I deliberately kept the enrollment small. In subsequent semesters, after the technicalities had been worked out and the procedures tested, the class could be enlarged. Now, however, it would be premature to have many in the class, because problems are bound to occur in any new venture, and particularly one without adequate funding. I warned prospective students, verbally and in writing, about the uncertainties, and stated that I wanted only adaptable, intellectually capable people. The four students who enrolled lived up to these high expectations. Ironically, none was a sociology major or had much sociology background. The three undergraduates were in journalism, astronomy, and history, while the graduate student was a political science major.

The project was centered on the Fall, 1980 semester. We met each Tuesday afternoon for two and one-half hours.

Connecting the Classes

Although direct telephone contact between the California and Hawaii classrooms would be the simplest and surest method, the cost of 40 hours of class time would be prohibitive—about $1,500. Therefore, I proposed to use the free PEACESAT radio service between here and the mainland terminal at Santa Cruz, and from there connect by telephone with Northridge. This would cut direct communication cost by more than half, to $683.

My request for funds for this purpose was rejected by the three major potential sources of such support at UH: the Graduate Research Council, the Educational Improvement Fund, and the UH Foundation. However, a direct appeal to the Office of Research Administration succeeded in getting a few hundred dollars—not enough for the link itself but at least enough to cover some incidental expenses, such as equipment, texts, and supplies.

A further problem emerged during the summer: it became apparent that PEACESAT would not be available for this project, for several reasons not related to technology.

Fortunately, I had assumed that there would be hitches in any new enterprise, so I had already started searching for back-up systems six months before the course was to begin. These possible alternatives ranged from amateur radio to computer conferencing, direct radio linkages to NASA Ames Research Center or Jet Propulsion Laboratories, the federal telephone service, and private or government WATS lines. A WATS line is a long-distance telephone service for which a flat rate is charged for an almost unlimited number of calls. Actually, there are layers, so that the first ten hours cost $326, with lower rates for additional time until, after 200 hours, the cost would be only $4.00 per hour.

In addition to a number of individuals and departments within the University, I also contacted outside agencies, including the Army, Navy, Air Force, National Weather Service, Federal Aviation Agency, Civil Defense, Congressmen, regional and national offices of NASA, the State of Hawaii Department of Planning and Economic Development, the National Science Foundation, and Hawaiian Telephone Company.

I knew before I started that there would be surprises and disappointments, and I was right. The biggest disappointments were the rejections by organizations I thought would be helpful: the aerospace program at UH, the Air Force, Jet Propulsion Laboratory, and most of all, NASA. One reason why I had selected aerospace as the subject for this experimental class was that this topic has a wide appeal, and I thought that more support would be available for it than for a course in art, philosophy, or the traditional topics within sociology. A NASA public education officer at Ames Research Center tried to help but was blocked by regulations and apathy.

Another disappointment was that the time at which the course started was not an advantage. 2:30 p.m. in Hawaii was 4:30 on the West Coast and 7:30 p.m. back on the East Coast, after work hours for the mainland, so there would be plenty of unused communications capacity. No one, however, offered the use of their lines after hours.

A third disappointment was that appeals in the name of science, education, or public service were unsuccessful. I had hoped that appeals to at least one of these worthy causes would motivate someone to help us, in return for favorable publicity or tax benefits, but it did not work out.

A month before the fall semester began, I did find an enthusiastic communications specialist in a federal agency. He offered to arrange use of a government telephone line as an educational and public service. We pretested the connection, calling from his office to the main federal switchboard in Hawaii, which connected us with headquarters in Washington, D.C., which in turn connected us with the professor in southern California. Success—or so it seemed. When the actual day of the class arrived in September, we pretested the procedure again, several hours before the class was to begin. This time a federal operator refused to place his call, on the technicality that the UH campus, or even his federal office, was off the federal network and therefore could not go to another "off net" number, the California university. It would be all right if we originated the call from the federal building in downtown Honolulu, but we were unable in the ensuing weeks to find anyone in the Federal building who would allow us to use their phone.

So the course began without the mainland link. Every day I spent an hour or two trying to locate funds or a free communications link. As the weeks went by without contact with California, our class proceeded on its own, sometimes listening to cassette tapes of the previous week's class in California. After several weeks we remarked, only partly in jest, that even if we did get the contact,

"they" could probably learn as much from us as we could from them. One of our members developed a space flight simulation exercise, many helpful articles and reports were brought in and shared, we had some stimulating conversations, and got intriguing insights from a UH anthropologist who had sailed a replica of an early Polynesian ship to Tahiti without modern instruments; his observations on group dynamics during extended voyages were very relevant to space flight.

Finally, help arrived from a most unexpected source. The athletic department offered us the use of its WATS line. There were several extension phones tied onto it, so several of us could participate at the same time. And, we could use them for a couple of hours each week!

Continuing Concerns

We were happy to have the link, but there continued to be frustrations. Being dependent upon someone who was very busy and had many responsibilities raised the concern that an emergency might preempt our access to the phone. This did not happen, but it made me aware of the importance of predictability and control in a telecommunications class.

A more immediate problem was technical quality. It was hard to hear what was going on at Northridge. We started with speakerphones purchased from Radio Shack, but conditions in both classrooms rendered them unsatisfactory. The California classroom was large and noisy, making it difficult to pick up voices more than three or four feet from the phone. We therefore rented a more sensitive speakerphone from the southern California telephone company. In Hawaii the room and the class were smaller, but there was background noise and the wiring was too complex for us to install a plug for our Radio Shack instrument, so here too we had to have a speaker attachment installed by Hawaiian Telephone Company, so we could all listen at once.

These instruments, while better than the Radio Shack equipment, still did not pick up all the California voices adequately and did not present them here loudly enough to be easily heard over the noise of nearby construction, voices in adjacent rooms, and passing vehicles. Even the air conditioner added noticeably to the din. Often, we could only hear snatches of conversation.

In spite of the frustrations, we heard enough to make it worthwhile, so we listened for parts of four weekly sessions.

For the last day of class, I arranged a conversation with the most prominent specialist on the social aspects of space: Dr. Stephen Cheston, Dean of the Graduate School at Georgetown University.

Although it was only mid-afternoon in Hawaii, it was evening in Washington, so we reached him at his home. He answered our questions and gave us an update on current developments in the space program, including information from conversations he had that same day with key figures in the nation's capital.

This was an ideal use of telecommunications for such a class. The voice quality was excellent and the discussion went perfectly, enabling us to end the course on a very satisfying note after all the frustrations of preceding weeks.

Discussion and Conclusions

We return now to the questions raised at the beginning of this article: *Can* it be done? Can two-way telecommunications, without adequate funding and facilities, link classrooms thousands of miles apart? Yes.

Should it be done?

The answer depends on our objectives and on the amount of support available. Low-budget telecommunications can be a stimulating supplement to an ongoing course, but it is too unpredictable to be the main component of any class. For a genuinely integrated class, in which students depend for their direction upon someone far away, adequate funding is necessary. Otherwise, too much time and energy will be wasted, both for teachers and for students.

A familiar analogy is the use of movies in regular classes. Teachers from kindergarten through college have experienced the frustrations of ordering a film, getting a projector, having a room full of pupils ready, and then encountering trouble with the machine.

Because telecommunications is more complex, there are more possibilities of things going wrong, and it only takes one minor malfunction to shut down an entire system. Therefore, teachers should be judicious about trying it. Under the right circumstances, telecommunications can be a marvelous educational aid, but unless a telecommunications system is in excellent condition—and this includes personnel and procedures as well as equipment—it would be wisest to limit its use to small classes of mature students or to occasional enrichment activities for a larger class.

Long-distance learning has been with us for decades in such forms as correspondence courses conducted through the mail. Modern technology has expanded the possibilities to the point at which teachers and students anywhere in the world *can* communicate instantly with each other. The technology is available now, at a cost within reach of many educational institutions. It is a pity that more people do not have a chance to *use* it. □

Electronic Mail: An Exemplar of Computer Use in Education

Heinz V. Dreher

Introduction

It is by now well established that the computer industry in general is advancing at near breakneck speed. This poses special problems for users of the technology, including users in education. Caution needs to be exercised so that we don't expend valuable resources for trivial gain. People, for example, are superior to computers in adapting to new and constantly changing requirements so often found in teaching situations. In these circumstances, it may make better sense to use computers in *facilitating* and *supporting* the teaching role as opposed to replacing it. This article addresses the problem of making appropriate use of computers in educational settings. *Electronic Mail* provides an example of such use.

Appropriate Use of Appropriate Software

Good software tools are expensive, must be implementable on available hardware, and need to be complemented by adequate documentation. Since the education sector typically has a paucity of funds, there has been a tendency to opt for what may be called simplistic computer exposure. This is particularly noticeable at the primary and secondary levels—witness the number of schools in which programming is offered. All computers, however small and cheap, can be used to "teach programming." Problems arise from the fact that most teachers are undisciplined programmers who create, and teach their students to create, unstructured or "spaghetti" code. The implications of writing such code in the development of commercial software and in its subsequent maintenance are enormous, as most software people will gladly testify. Such approaches can also be very frustrating for students, in addition to being educationally futile.

Heinz Dreher is a Lecturer in Information Processing attached to the Western Australian Institute of Technology's School of Computing and Quantitative Studies, Perth, Western Australia.

Computers in Education

Often computers and education are associated via CAL (Computer-Assisted Learning) and variations thereof (Computer-Assisted Instruction, and Computer-Managed Learning). Then there is the traditional link with mathematics. It is true that computers were born out of man's need to crunch numbers, but mathematicians form a very small minority of computer users now. Doctors, police, journalists, and authors, and of course the business and industry people, are prominent among today's users. They have essentially three fundamental requirements in common. One is to search and shuffle data or information; the next is to communicate; and the third is do it all quickly, accurately, and reliably. Relatively few people have the need to calculate by computer, and probably fewer need it to assist learning in the CAL/CAI sense.

While there is no doubt about the advantage that may be gained from using computers in these typical educational roles, it may be constructive to envisage some new and different areas of applicability in education.

Seymour Papert[1] sees the computer being used in new and novel ways.

> ... "computer as pencil," that is to say, emulating the quality of the pencil as a familiar, freely available object that can be used for many purposes, such as writing, drawing, scribbling, doodling, calculating, chewing, and so on.

This opens a wealth of applications of the technology, and all within education too. Students could use word processors for their assignment work, or statistical software to remove the tedium of manual calculations. Graphics capabilities of modern computers provide endless opportunities. The emphasis is on the student using the computer, in a way meaningful and useful to him, in *supportive* and even *management* roles.

We like to think of computers as providing an extension of an individual's own capability and capacity, rather than just a substitute for a particular manual role.

Computers and Communication in an Educational Setting

Communication is a vital component of our daily lives. Mostly our communication is verbal, and when a more permanent record is required we usually resort to writing. Until recently, this meant the use of paper and pencil. Now, with the proliferation of affordable word and text processing systems, it is becoming more feasible to implement paperless environments. The computer industry caters admirably to our need to communicate.

One avenue that we have been exploring at the

Western Australian Institute of Technology is to make use of the excellent software originally provided for the business sector and use it in an educational setting. We use the computer to facilitate and improve communication between students and faculty. Assignments are prepared in electronic form using a text editor or word processor, submitted for assessment and review via *electronic mail*, and subsequently returned to the originator, again via electronic mail, of course. In this context, we focus on facilitation of learning through *support* rather than *instruction*.

It is suggested that our use of electronic mail in a support role is educationally appropriate. Students need to communicate, and are familiar with sending and receiving ordinary mail. Making the reasonable assumption that the chosen electronic mail package closely resembles the familiar postal system, students require but little guidance in sending and receiving electronic mail. Consequently, they become free to concentrate on deriving the benefit accruing from its use, as opposed to learning about yet another new and different scheme. Benefits are numerous and include simplicity of creation, editing, and submission of assignments. No longer can there be doubt about the lateness of submissions (in case it's of importance) as the *electronic envelope* contains an *electronic postmark* which becomes an integral part of the letter once it is *posted*. Neat, type-written hard-copy can easily be generated should it be required. Multiple copies of the electronic document are easily shared, which in our case is a privileged and restricted function for obvious reasons.

The emphasis is on ends rather than means. This combination of a familiar scheme (postal system) together with the very *utility* of that scheme, makes an electronic mail package eminently suitable as an instance of using the technology in an educational setting—in our case, at first year level of a Bachelor's Degree course in Business.

An Appropriate Software Package: MAIL

So, we have made a choice, and it is implemented on available hardware. The package MAIL[2] resides on a DecSystem10[3] and is adequately representative of the usual mail system. This product was made available late in 1982, at which time it was used experimentally with a small group of first year students. A favorable student response added encouragement to the plan to make wider use of MAIL during 1983.

MAIL "knows of" all users of the computer system; that is, everyone has an *electronic mail address* to which *mail* may be sent by other users. Each letter is *postmarked* with date and time, and then forwarded to the addressee. On the next occasion the addressee is connected to the computer system, he is informed of the presence of *mail* in his *electronic postbox* and may read it or deal with it in a way analogous to the conventional system. One may reply, file, forward, or even postpone any action until some specified future time. If a user has a requirement to remind himself of a commitment, he simply sends *mail* to *himself* to arrive in time to act as the required prompt.

There are many more useful features besides. MAIL can keep track of all correspondence, including replies and forwarded mail items, which is very useful for subsequent follow-up or documentation purposes. Sending copies of correspondence is also catered to. One of the most useful capabilities of MAIL, for instructors at least, is the ease of maintaining, and sending *mail* to, address lists. Thus, instantaneous "broadcasting" is little more difficult than sending one letter only. In cases where it is important to know whether *mail* has been received (and even read) by the addressee, the required information can be readily obtained by making inquiries of the *electronic post office*.

Students' Views on Use of MAIL

Students have found MAIL beneficial as indicated by strongly positive responses to a survey conducted toward the end of the semester. First, in relation to the use of MAIL for assignment-related work, students reported that they had used it to advantage. Significant benefits accrue from the simplicity of submitting assignments for assessment and from the subsequent return of feedback augmented with samples of "good" and "not so good" assignments. Time previously spent by an assessor shuffling, copying, and comparing paper documents could now be re-directed to providing constructive feedback. The task of keying in assignment text was, for the majority of students, not an excessive burden.

A second, and very positive, effect was that an instructor was suddenly found to be but a few keystrokes distant, as compared to the more usual situation of being inaccessible in a remote office, with all manner of protocols required to gain his attention. Particularly for requests requiring short answers, clarification of procedural matters, and even making appointments in case of problems best dealt with in a face-to-face situation, MAIL was a real boon. And if a student query with its attendant solution was worthy of sharing within a wider student group, *bulk mail* could be readily sent there and then, as opposed to waiting until the next class meeting.

Quite obviously, it is not meaningful to consider using MAIL without first establishing adequate access to an appropriate computing resource. For us that meant 120 odd students having twice daily terminal and DecSystem10 availability, a not inconsiderable demand. The actual load was much greater than anticipated, substantially due to students communicating among themselves while the novelty factor was still present. At times so much *mail* was being processed that *electronic letter boxes* overflowed—with unpredictable results, but solidly reinforcing the need to manage carefully one's own affairs in case of catastrophe.

Instructor Reaction

While only a minority of students reported the use of MAIL as problematic, the majority of instructors working on the team expressed concern. Comments such as

> it concerns me that the submission of assignments for this unit will apparently be done using the MAIL system . . .

were prevalent among tutors at semester start. Perhaps it comes as no surprise that students adapted to the change far more readily than their (older) instructors, who were essentially concerned with the changes to their routine due to the perceived extra demands the use of MAIL would make on their time. An oft-repeated justification for averting extensive use of MAIL was the relative unavailability of computer terminals. In actual fact this lack of adequate instructor access to terminals was a serious matter, and one that should have been foreseen.

The increased accessibility of instructors to students did create extra workload, but with hindsight this tendency may be ameliorated by the publicizing of guidelines aimed at self-regulating over-use.

What Has Been Achieved?

Although we are still some distance short of implementing a completely paperless environment (which may not be entirely desirable, in any case) we have succeeded in using a thoroughly up-to-date approach to communicating in our educational setting. Since graduates from our course enter business and commerce, where such tools have been in use for some time, valuable preparatory pre-employment experience has been gained.

Not only is the approach modern, it is efficient and effective too—that's why the business sector *uses* it. Students themselves, in a comprehensive survey at semester end, gave broad approval and judged the benefits to be positive.

Let us recap on the benefits. We have built upon the student's prior knowledge and experience; exposed him to an up-to-date communication software environment with MAIL; facilitated learning through automated assignment creation, sub-mission, and feedback dissemination procedures; and gently immersed him in a computing environment so that he may feel comfortable in further exploring its potential. There have been benefits for instructors too: no long queues of students waiting outside their rooms or insisting on attention at inconvenient moments; simplicity of sharing information with a wider student group; absence of arguments in relation to misplaced assignment submissions; and very importantly an opportunity to document thoroughly the entire educational "goings on."

Finally, for those in our midst who perhaps tend to take the accountability issue in education to extremes, a word on costs. While it may be true that currently an electronic document costs perhaps three times that of its paper counterpart to distribute to students, pure "money sums" are far too coarse a measure on which to base final judgment. For the author at least, the future course is clear—the benefits are just too real and valuable.   □

Notes

1. Seymour Papert. Redefining Childhood: The Computer Presence as an Experiment in Developmental Psychology. In S.H. Lavington (Ed.), *Information Processing 80*. Amsterdam: North-Holland Publishing Co., 1980, p. 993.

2. *DecSystem10 Disk File Doc:Mail.Doc, W A Institute of Technology*, contains a user manual for MAIL. Also "help" files exist and may be invoked either at monitor level or from within MAIL itself.

3. The DecSystem10 is a large timesharing computer system capable of running a hundred or so jobs simultaneously. Another suitable hardware configuration would be a network of personal computers, connected either via a LAN (Local Area Network) or a public facility such as Ausinet in Australia. This is a cost-effective option considered by many schools throughout this country and elsewhere.

Acknowledgments

The author wishes to thank the contributions of students and faculty associated with Programming 103 in the first semester of 1983. Particular thanks to my colleagues R. Hodgson and R.D. Galliers.

MAIL was not written at WAIT and had its origin at the National Institutes of Health, Bethesda, Maryland USA, with Glenn Richart as author. MAIL has been extensively modified, however, to suit our particular use in a teaching environment. For example, the sending of the same letter to an entire group of students (an assignment specification, say), was problematic in the original version. Bulk postage, so to speak, has now been implemented. Further extensive modifications to MAIL, making it optionally menu driven, incorporation of a text processing module, generally improving the user-interface and enhancing functionality, are in the development pipeline.

Tailoring Telecommunications Innovations to Fit Educational Environments

Judith B. Harris

Will computer telecommunications innovations be successfully adopted and productively applied in educational environments? The answers to this question depend, in large part, on innovation attributes and their functional "fit" to K-12 and university cultures. The Curry School of Education at the University of Virginia and IBM's ACIS division launched a joint project in January of 1987 to develop diffusion methods for adoption of computer-assisted telecommunication technologies. Some of the most interesting interim results that have emerged concern electronic network attributes that can either encourage or discourage the development of an electronic educational culture. The purpose of this article is to present these findings as a way to explore the utility of a popular model of innovation diffusion that can inform the development and design of educational computer telecommunication technologies.

Theoretical Interpretations

Everett Rogers (1983) defined **diffusion** as "...the process by which an innovation is communicated through certain channels over time among the members of a social system." (p. 5) Rogers describes **communication** as a two-way exchange of information, the aim of which is to achieve mutual understanding about a new idea. He goes on to define **technology** as any design for action (idea, method or object) that reduces uncertainty in the cause-effect relationships inherent in achieving desired outcomes. Rogers' longitudinal work about the process of adopting innovations, long considered to be the standard by which diffusion efforts are interpreted in many disciplines, will serve as the theoretical lens through which some interim results of this project will be focused.

Judith B. Harris is Training and Field Coordinator, Teacher-LINK, Curry School of Education, University of Virginia, Charlottesville.

Rogers' cross-disciplinary and cross-cultural study of innovation diffusion yielded six factors that influence the success or failure of such efforts. Potential innovation adopters must:

1. perceive the new technology as having **relative advantage**; that is, being better suited to accomplish usage goals than the existing technology;
2. perceive the new technology as being more **compatible** with existing usage conditions and goals than the existing technology;
3. perceive the new technology to be less **complex** than existing methods for accomplishing similar goals;
4. perceive the new technology to have **trialability**; its use can be introduced gradually, allowing time for acclimation to new usage patterns;
5. be able to easily **observe** the beneficial effects of adopting the new technology; and
6. Finally, the new technology's inherent structure must encourage innovation adopters to independently **re-invent**, or design new uses for it. (Rogers, 1983)

The Project

During the early spring of 1986, several professors at the Curry School of Education at the University of Virginia made an enticing discovery: a cluster of minicomputers at the University's computing center offered an easy-to-use electronic mail system which could be accessed at any time of the day or night with a toll-free telephone call. With the professors' encouragement, several local classroom teachers, school administrators, university professors, and graduate students who were already collaborating on instructional computing projects (such as LOGO and adapting microcomputer use for physically challenged students) requested accounts on the system. They were welcomed by the university's academic computing user services staff, and found the "E-mail" to be facilitative to their collaborative efforts, if not downright addicting.

As more people heard about the convenient communications system, more people requested and established accounts. Users helped each other to acquire and configure modems for their personal computers, locate user-friendly telecommunications software, become familiar with the mail system's commands, and electronically send and receive word-processed files. The more people who joined the network, the more uses for the system emerged and were shared. Professors began to establish accounts for their students, using them to answer questions, send announcements, and provide assignment feedback. Committees began to send meeting notices and raise issues for discussion electronically, instead of on paper. University students provided practical and emotional support for each other with humorous messages, plans for get-togethers, and informative notes. This replaced "While-You-Were-Out slip telephone tag," since users could log on at any time of the day or night to send and receive messages at their personal convenience. It seemed obvious that this non-intrusive, easy-to-use messaging system filled a communications need that was not perceived until the technology became accessible.

Two professors who were eventually to become directors for **Teacher-LINK**, the cooperative Education School/IBM project, observed this grassroots diffusion process with growing excitement. If such a network could spread quickly and be found useful without deliberate infusion efforts, what then might be cultivated with planned support? More pointedly, might this new type of communication link K-12 teachers, long isolated behind the doors of their individual classrooms, with university staff and student teachers?

In late 1986, IBM agreed to supply the Curry School with approximately one million dollars' worth of computer hardware and software to help the professors explore the possibilities and practicalities of establishing and maintaining such a teacher telecommunications network. IBM wanted to know if an equipment and training package designed to assist schools of education in creating and supporting such a network was feasible, saleable, and profitable. The Curry School wanted to explore how to establish such an electronic educational communications culture. Both were curious as to whether use of an electronic teacher network would improve the quality of student teaching internships, bolster communication between universities and public schools, and encourage instructional idea sharing among teachers in different schools.

Early in 1987, an IBM 4361 mainframe running the VM/CMS operating system and offering an IBM electronic mail and messaging system designed for professional corporate office use (PROFS) was installed in the university's academic computing center. PROFS was substantially different from the UVA-authored minicomputer mail system ("Virginia Mail") that already supported approximately 40 teachers from the University and local schools, but since all UVA mainframes were now connected to a local area computer network (LAN), messages could be sent between different mainframes and minis without forcing users to change mail systems.

Five English teachers from local public schools and their student teacher interns were loaned portable IBM computers with 1200-baud built-in modems, printers, and IBM personal productivity software (word processors, databases, spreadsheets, and other such tools). Since informal adoption of electronic mail had proven to be so easy with Virginia Mail users to date, one two-hour training session was scheduled for all initial participants. Unfortunately, the microcomputer telecommunications software failed to operate, frustrating both project directors and participants. A licensed shareware product was quickly located to substitute, but not being designed to function well with full-screen displays transmitted at 1200 baud, it proved to be at best a threadbare "patch." Teachers found using a screen-oriented E-mail system originally designed for high-speed terminal use on a 1200-baud personal computer to be discouragingly slow and non-intuitive. Eventually, the university's academic computing programmer who was assigned to facilitate the project was able to massage the mail system into semi-friendly functioning capacity, but only after two of the original five frustrated supervising teachers had voluntarily left the project.

It therefore became apparent to the project directors that this E-mail system required more user support than the Virginia version, which had spread and was continuing to proliferate with minimal, if any, formal training efforts. They requested and received a graduate student quarter-time position to provide the needed network use facilitation. After one year of modest success, the dean agreed to double funding for project support when it became apparent that twice the allocated hours were required to keep an increasing number of network users comfortable with this new form of communication. Three additional part-time positions were funded to assist professors with the infusion of educational technologies (including, but not limited to, electronic network use) into instruction at the school of education. Interestingly, most of the support supplied for E-mail use was directed toward PROFS users, as user needs dictated.

As the semesters passed, more supervising teachers ("clinical instructors" in UVA parlance), student teachers ("student interns"), professors, and supervisors were supplied with hardware and software and introduced to the network. Five Social Studies and five Elementary Education instructor-intern pairs were welcomed in the fall of 1987, five mathematics, five science, and five more elementary education duos were added during the spring 1988 semester, and three special education and two physical education pairs were included in fall, 1988.

At the same time, user numbers on the Virginia

mail system were growing by leaps and bounds; plans were approved and implemented in the fall of 1987 to have all undergraduate education students online by the fall of 1989. Approximately 80 percent of the professors and 50 percent of the graduate students had E-mail accounts by January 1989, and more than 60 local teachers and administrators had requested, received, and were using Virginia Mail accounts by the same time. The UVA-authored system maintained its popularity over PROFS, despite the fact that PROFS users were loaned equipment and software, given 6-8 hours of training, 24-hour support, and telephone lines into their classrooms. In contrast, most of the Virginia Mail users had to locate hardware, telephone connections, telecommunications software, and occasional user support on their own. While approximately 33 percent of the PROFS users became active participants, at least twice that proportion of Virginia Mail account owners used the network regularly.

During the last months of 1988 and the early months of 1989, when the full-scale phase of Teacher-LINK (80 PROFS users and several hundred Virginia Mail users) was being planned and executed, it became apparent that sluggish adoption of PROFS as compared to the rocket-fueled embracing of Virginia Mail could quite possibly be caused by different levels of user interface friendliness. System documentation, training, and support strategies had been developed and refined to maximal efficiency during the first two years of the project; participant selection was limited to only those teachers and interns who expressed the most convincing motivation and time commitment for testing the technology. Yet PROFS still drew complaints from users, while Virginia Mail users proliferated relatively unsupported.

Teacher-LINK staff, though tempted to switch all users to the UVA-authored system, decided to try another E-mail system that could run on the IBM 4361 mainframe in a final effort to provide IBM with encouraging news about the diffusibility of an IBM mainframe-based teacher telecommunications package. At the present time, this system ("Rice Mailer") is being user-tested with the final group of Teacher-LINKers. The final verdict will be clear by January of 1990, when the newest group of participants will have used the network for 10 months, four of which will comprise the student internship. In the meantime, IBM is supporting the development of a menu-driven mainframe user interface with its own easy-to-use E-mail system. Developed and field-tested by personnel at Iona College in New York state, this package ("ConnectPac") shows promise for university and

K-12 teacher networks principally because it makes mainframe functions seem like personal computer DOS commands. Users can even configure their accounts so that their favorite word processor is used to generate all electronic communication, cushioned by simple menu options that help users send, receive, and organize word-processed messages and files.

Contrasting Project Perspectives

Simon Fraser. A similar project began coincidentally in 1986, more than 3000 miles away at the school of education at Simon Fraser University (SFU) in Vancouver, British Columbia. Neither group was fully aware of the other's efforts until a chance encounter occurred in late 1988 between two facilitators through BitNet, an international telecommunications network of colleges and universities which was available to all participants in both projects. The Canadian project (**XChange**) had about 3200 users online by early 1989, 1300 of whom were described as active participants.

Simon Fraser teachers used a conferencing system primarily; secondarily an E-mail system. Virginia teachers used the reverse configuration of electronic communications structures until early 1989, when Teacher-LINKers were formally introduced to electronic conferencing as part of their new member workshops. Networked teachers' usage statistics then began to indicate a clear preference for the conferencing system (*Caucus*), which was donated to the project by MetaSystems, Inc. Participants described *Caucus* as easier to use than PROFS. Project staff noticed that public forums for discussion (achieved through the use of conferencing software) seemed to promote information exchange and facilitate professional and personal support structures. User-friendliness and public discussion consequently encouraged more frequent use of the network. It seemed that the hoped-for "electronic culture" had established itself by April of 1989.

SFU had also developed and revised documentation and training to streamlined forms by early 1989, but was not experiencing the difficulties which confronted Virginia's PROFS users at that time. David Bell, SFU's project coordinator, explained at a conference presentation in February 1989 that when professors at his school discovered the electronic mail capabilities of SFU's mainframes (IBM machines) and began to experiment with the system, they quickly realized that the user interface must be changed to a simpler format before novice users could use the network comfortably. The school of education hired a full-time programmer to develop, install, and support the FFE ("Friendly Front End") that would accomplish this goal. With clear documentation, user-considerate online prompts, and the help of the equivalent of four full-time support personnel, XChange is serving its networked teachers' communications needs quite well. This success seems to be supported in large measure by the FFE itself, an aspect of UVA's PROFS system that was lacking.

IBM Research Laboratory. Denise Murray's 1986 study of language use in electronic mail, messages, and conferencing in an IBM research laboratory provides another interesting contrast to the UVA network usage profile. These IBM employees were far from novice computer users, and made frequent, almost effortless use of the PROFS system. The same user interface that bordered on technical opacity for UVA's teachers was close to functional transparency for them. Yet both groups engaged primarily in what Murray terms "conversation for action," with supportive, often humorous interchange buttressing the action centered intercourse. Murray examined how the IBM employees' choices of communications medium (paper, face-to-face, or computer-assisted) and mode (i.e., computer mail vs. conferencing) corresponded to content of different interchanges. She concluded that these choices were far from arbitrary; instead, they were selected depending upon the contexts of information exchange situations from a full range of communications options. If one or more of these options were less accessible than the others, due to an inconsistent match between user facility and front end user accessibility, then it would seem that her informants, like the Virginia teachers, would have made inconsistent communications medium choices.

National Urban League. In early 1987, headquarters for the National Urban League in New York City began a three-step adoption of a national electronic mail system. The diffusion plan was carefully crafted in top-down format, much like that of the IBM installation. Administrative superiors in headquarters, regional and affiliate offices were trained in that order to use the system, which now replaces much of inter- and intra-office paper communication. Several NUL employees were already using computer equipment to collect and interpret data, write documents, and create presentation materials; few, if any, had ever used electronic computer communication systems. Most described themselves as relatively inexperienced computer users, much like the teachers in Virginia and British Columbia. Use of the Urban League system was a job requirement, much like network users in Murray's regional IBM office, and in direct contrast to the two voluntary university teacher telecommunications networks.

Interestingly, IBM offered the Urban League an identical system to the one accepted by UVA professors: a 4361 mainframe equipped with PROFS. NUL officials declined the offer after experimenting with use and maintenance of the system, accepting instead a one-year partnership with Digital Equipment Corporation, which supplied a standard integrated function user interface that satisfied NUL's adopter requirements. Personal conversation with NUL's systems analyst/PC coordinator yielded an appreciable amount of sympathy for UVA's predicament with PROFS; NUL's employee labelled it "a bear."

Within one year, all administrative personnel in all 119 Urban League offices across the country were trained to use the system, and they continue to do so. Kenneth Small, executive assistant for strategic planning for NUL, attributes this successful diffusion primarily to the fact that information that is perceived to be needed and valuable by NUL employees is exchanged expediently through the network. Perhaps it is also true that both the IBM and the NUL networks succeeded in part due to use being one aspect of a stated job responsibility, as opposed to use being a volunteer effort in the two electronic teacher communities. It is interesting to note that the diffusion strategies adopted in the two commercial organizations were decidedly hierarchical and implemented from the top down, while the infusion patterns for both public school/university networks were non-hierarchical ("grass roots") and operated from the bottom up. Yet despite organizational differences, the match between user characteristics and comfort with different network interfaces in all situations seems to be the best predictor of successful innovation adoption.

Conclusion and Implications

The data from these four studies can be interpreted to suggest that the structure of communications mode must be carefully fitted to the preferences, needs, and prior medium experience of projected innovation adopters if innovation diffusion is to be successfully accomplished. This hypothesis will be further substantiated if the change of mail system in UVA's 4361 mainframe results in easier and more frequent use of the Teacher-LINK network.

Stated in terms of Rogers' innovation attributes, the following interim results are suggested from project work to date.

- **Relative advantage:** In all cases cited above, users found electronic communication to be more convenient, less intrusive, less expensive, and more expedient than "telephone tag."
- **Compatibility** with existing usage conditions

and goals: All users except those attempting to implement PROFS in Virginia found electronic mail more helpful in carrying on essential communications than telephone or paper met methods.

- **Complexity**: Although it is unclear whether any of the E-mail or conferencing systems in the four cited cases were perceived to be less complex than using the telephone or paper and pencil, it seems that all systems but PROFS (at UVA) were not complex enough to discourage regular use.

- **Trialability**, or gradual introduction of innovation use, allowing time for acclimation to new usage patterns: Certainly an alternate form of communication, when three are already in use (paper, telephone and face-to-face communication) boasts that advantage, even if the user interface is unfriendly.

- **Observability**: Perhaps one of the best forms of "conspicuous consumption" occurs when a computer user displays his/her hardware in the office or classroom, and shares the information and experiences gained through use of the new technology with peers. It seemed apparent that all network users in the four instances cited did this in proportion to their comfort with and excitement about the benefits of using the innovations.

- **Re-invention**: The proliferation of creative use ideas in all contexts mentioned, most notably among networked teachers, attests to electronic networking's potential in this area, despite interface friendliness.

Computer-assisted telecommunications can offer teachers at K-12 and university levels increased communications convenience, easier and less expensive access to information, and expedient and frequent communications links with previously inaccessible peers, thus lessening the commonly experienced sense of professional isolation that is a pervasive characteristic of instructional occupations. With streamlined provision of equipment, software and system support, teachers can easily and permanently appropriate this new technology, adapting it to their individual instructional and management needs.

Yet if a single element of this multi-faceted mixture of machines, mentors, and motivations fails to expediently adapt to the needs or styles of potential networked communicators, multiple adoptions can be irrevocably threatened. Seeming details such as the structure of a user interface can explode into full rationales for innovation diffusion failure. Conversely, user individualized and situation aligned equipment, systems, and support mechanisms can virtually guarantee quick and longstanding innovation adoption, despite even less than ideal machine availability or user support mechanisms. If it is our goal to modernize communications modes in education, we would be well advised to design and implement these with teachers in mind and alongside. □

References

Edghill, M. *AEA Network Presentation.* Paper presented at the meeting of the American Evaluation Association, New Orleans, October 1988.

Lomax, F. *Consensus Building, Evaluation Methods, and the Computer Based Network: A Case Study of the National Urban League.* Paper presented at the meeting of the American Evaluation Association, New Orleans, October 1988.

Murray, D.E. Conversation for Action: The Computer Terminal as Medium of Communication (Doctoral Dissertation, Stanford University, 1986). *Dissertation Abstracts International*, 1986, *47*, 4702A.

Porter, D. *XChange Communications Guide, version 2.1.* (Available from Nola Jones, Faculty of Education, Simon Fraser University, Burnaby, B.C. V5A 1S6.) October 1988.

Rogers, E.M. *Diffusion of Innovations* (3rd ed.). New York: The Free Press, 1983.

Computer Mediated Communication for Instruction: Using e-Mail as a Seminar

Alexander J. Romiszowski and Johan A. de Haas

Computer mediated communication (CMC) is a very fascinating instrument for both distance education and the conventional instructional methods (so-called "place based" education—Levinson, 1989). Our interest in CMC was raised by the Conference on CMC in Distance Education held at the British Open University in October 1988. It presented the O.U.'s first year experience with the use of CMC as a medium in distance education courses (Bates, 1988; Mason, 1988). In this article we would like to describe some possible applications of CMC and especially the use of computer conferencing systems. Focusing on the educational value of CMC, we next try to give some suggestions to be taken in consideration, when using CMC as an instructional system. In the final section we present some experiences we had conducting CMC between the USA, Canada, and the Netherlands.

Computer Mediated Communication: Overview Computer Conferencing (CC)

CMC has principally been used as a conference system, hence the popular term "cc." This implies that a group of individuals exists, who choose to debate a topic of mutual interest. They may be in search of new ideas and may view a change in the direction of the discourse as a positive development. Also, individuals may join in or drop out of the discourse, in relation to their own interests in the main topics under discussion. The continuance of the discussion is controlled by the existence of a "critical mass" of interested discussants.

The marriage between computers and telecommunication made computer conferencing possible from 1970 onward (Toner, 1983). However, within the social sciences, the first computer conference was not conducted before 1982 (by the Western

Alexander J. Romiszowski, Professor at Syracuse University, New York, is a Contributing Editor of this magazine. Johan A. de Haas is Assistant Professor at the University of Twente, Division of Instrumentation Technology, The Netherlands.

Behavioral Science Institute, La Jolla, USA). Before this, computer conferencing systems were only a matter of business and technology (Katz et al., 1987). Within training it was mostly used in science and not in the social studies (Feenberg, 1987). Feenberg and Eldridge (1982) stress the importance of cc, because they feel the communicative infrastructure will change in this direction: ". . . the failure of the liberal arts to participate in the process of invention of these new computer based teaching technologies may result in a wider crisis of humanistic learning in the future, when it turns out that the technical infrastructure of the university had been reshaped exclusively in function of the need of scientific and professional education" (Feenberg, p. 119).

Computer conferencing systems are offering great opportunities for distance education. Lauzon and Moore (1989) envisage computer conferencing as a fourth generation of technological delivery systems in distance education. After correspondence courses, audio conferencing, computer based training for individual instruction, there is computer conferencing providing "the potential to deliver both asynchronous individualized instruction and group instruction to distance learners (p. 43).

To many persons computer conferencing systems are a kind of electronic mail which, besides this function, facilitate specific group communication tasks. Most of these systems have possibilities for electronic mail, search and/or organizing functions (branching), file transfer, and editors. The total amount of users who can be logged on at the same time changes with the conferencing system and the computer configuration one uses. For example the CAUCUS conferencing system on an IBM PC XT or AT model 30 can have only one user at the same time, but when used on a VAX 8000, CAUCUS can host 80 simultaneously logged on users. A few well known systems are:

EIES   Electronic Information Exchange System, developed by the New Jersey Institute of Technology in 1977.

PARTI   Abbreviation of "Participate," the first conferencing system to offer "branching." Parti is available on The Source, NWI and Unison.

CONFER   Abbreviation of "conference," developed by the University of Michigan, 1977.

CoSy   Conferencing System, developed by the University of Guelph, Canada.

These and other systems are used in a variety of courses. Here are some examples:

- The British Open University offered a course on the social and technical aspects of information technology to more than 1300 students using the CoSy conferencing system. The integration of the system is seen as a vital and exciting component for students to gain practical experience of the social and technological issues discussed in this course (Mason, 1988). This use of CMC as an integrated aspect of a year long course is possibly the biggest single educational application to date.

- Jutland Open University is offering courses in arts and archeology with cc. They had the experience that working on computers in special centers did not work, but having the possibility to work at home at your computer did. They view computer conferencing as a more democratic way of offering education (Lorentsen, 1988). This finding is particularly interesting given that the bulk of research on educational radio and TV is on group based study in special centers.
- Connected Education is a computer conferencing network using the EIES system. Since the fall of 1985 they have offered more than 100 courses entirely via computer conferencing for graduate and undergraduate credit in conjunction with the MA in Studies Program at the New School for Social Research and several programs at the Polytechnic University in New York City (Levinson, 1989).

Bulletin Boards

Another common use of CMC is as an information system, as exemplified by the proliferation of electronic bulletin boards. Their primary purpose is to offer a special interest group information sharing service. An example of this, in our own field, is the recently launched EDUCATIONAL TECHNOLOGY MAILING LIST. This was the idea of a group of graduate students and faculty at Michigan State University in February 1989. Having themselves been involved in the use of electronic mail for professional communication, they decided to set up a system of communication open to all persons interested in sharing information on educational technology topics. The mailing list was widely publicized in February. Starting from a (mainly Michigan State) membership of 17 people in February, the number of participants grew to nearly 100 in March and over 130 in April and about 180 members from 90 institutions in 15 different countries by the end of May 1989. Message intensity has increased from less than one a day on average in February, to over three a day in May. A variety of topics are being discussed and use of the system seems to be evolving from a bulletin board for announcing events and services, to a forum for discussion of current "hot" topics (for example, a recent particularly heated and prolific exchange on the question of copyright of material that is put out on freely accessible electronic bulletin boards) (Rosenberg and Banks, 1989).

Online Journals

Another way of using CMC for information dissemination is as online journals. This is an electronic equivalent to the conventional academic journal. It has the advantage of rapid publication and easy editing, updating, and also cooperative authorship of contributions by geographically separated authors. The above mentioned description of the EDTECH mailing list was published in the July 1989 issue of the *Online Journal of Distance Education and Communication* (Ohler, 1989, editor). Mark Rosenberg and Vickie Banks wrote their report no earlier than June (it quoted full data for May). It was published in the Online Journal on July 4th. And here we are quoting it in this article, writing on July 10th, and sent from The Netherlands to this magazine via facsimile transmission soon thereafter. This sort of rapid turn-around can be of great value in certain educational communication applications.

Social Communication

A further and growing use of CMC is as a social communication system. This use is of growing importance in distance education. It provides students with the opportunity of social interaction with other students, which often is a missing component in distance education. The experience of the British Open University showed that from the students' point of view, this was the most used and most valued function of CMC (Bates, 1988).

Our main interest is to examine the use of CMC as part of an instructional system. It is in this context that many of the advantages and problems discussed earlier are of greater importance and relevance to the designer of instructional systems.

Using CMC for Instruction: Benefits, Problems and Some Solutions

Computer conferencing systems are offering a "time and space independent" way of communication. These main characteristics are creating many positive expectancies towards the medium.

It provides geographically scattered people with a means to communicate. This can be of importance in distance education programs or in support of training (salesman traveling around the country, for example). Class can be open 24 hours a day and you never find a closed door when you want to speak to your tutor. It provides a more democratic environment for group interaction, in the sense that all people are equal because they have the same possibilities and tools to communicate their thoughts. Interruptions can be made at any time. Even reactions to past messages belong to the possibilities.

There is an increased potential for class interaction and this interaction can be more 'in depth,' because you can think over problems at your own pace and you have the possibility to look up other resources. A further benefit is that it provides everybody in class with a total record of the course. Most conferencing systems offer organizing structures that make it easy to split up in small groups or to discuss certain items in closed sessions. They provide opportunities for group learning and peer tutoring and have a particularly important social value in distance courses, where people normally lack contact with other students.

As for many things, highly positive features usually have also their darker sides.

The social equality of participants is indeed a feature of computer conferencing systems, but only in an initial phase. When the conference is going on, people tend to attach certain characteristics to messages, and this leads to differences in the status of the participants. Suppose a computer conference just started and you are reading the messages and every day you will find some very odd remarks from a Mr. de Haas. After a few days you will read this message with a certain skepticism or just skip his messages entirely. You can't say that the status of the participants in the conference is still the same!

The possibility to react at a later time may promote procrastination and in some cases failing to respond altogether. There are people who don't trust their thoughts in print. When you put in a message and nobody reacts, this tends to discourage further interest in the conference. In other words, there will be an amount of people only reading messages and never responding (so called 'lurkers').

A lot of problems you have when working in groups will come back in CMC. But the problems will be different, because we communicate through another medium. In a face-to-face meeting most people are anxious to know how others react to their messages. In CMC it is not different. However, CMC lacks even tacit signs of approval such as nods of the head, smiles, or glances, which are often immediate reactions in everyday life. Compare this with the distress it causes when you say 'hello' to someone and you don't get a reaction (which often happens in CMC), then you know why CMC amplifies certain social insecurities, which Feenberg calls "communication anxiety" (Feenberg, 1987).

Another difficulty with computer conferencing is the problem of getting a view of the structure of the discussion. It is very easy to drift away from the main theme of the conference. And it is always possible to react to a message way back in the conference. As a result this kind of discourse is not only "multi level" (several themes in discussion) but also "multi speed" (different aspects of a theme being addressed by different participants).

For the time being let us not think of computer conferencing as the new medium through which all problems will fade away. Too little research has been done in this area, although experiences with computer conferencing have already given us a lot of insights into possible solutions for the above mentioned problems of control of the process and overview of the structure of an evolving discussion.

CMC is a qualitatively different medium for instruction. When designing instruction for CMC, it is important to consider the strength of the medium and to focus attention to the learner.

A first thing to do is assure people that they actually can communicate with the system. You have to maintain a support structure to help solve technical problems. Next to this it is important to be sure people really do have access to terminals or personal computers and modems, so they can log in frequently.

To assure input of the participants, an active group leadership is necessary. This leader or moderator must be the host, setting a congenial, non-threatening climate, thanking people for their contributions, and stimulating them to react (again). But next to this he or she has to be a "chairperson": summarize the discussion, ask for clarifications, create unity, and watch the theme from drifting off track. And last but not least, the leader has to maintain the bunch of participants as a group. Group maintenance includes such duties as mediating differences that become obstructive and making comments pertaining to the group's progress.

Setting a climate helps to ease transition of students to both a new course and possibly a new learning environment. In order to accomplish this the instructor can use the following strategies (Feenberg, 1987; Davie, 1988; Romiszowski and Jost, 1989):

— leave a personal welcome message for each student;
— reinforce early attempts at participation;
— reference students' responses in your comments;
— send students individual (private) communication that provide feedback and suggest resource of possible relationships with other resources;
— model expected behavior, concentrating on content and thought provoking ideas, rather than such things as keyboarding skills and formatting.

The following are suggestions for being a group facilitator:

— keep the main discussion on track by providing leading questions;
— if the discussion starts getting off track, refocus;
— if a distracting topic appears that is generating interest, create a branch so that the competing conversation is separate but optional;
— focus effort by suggesting that students look deeper into topics when applicable;
— provide summaries of what has been transpiring by drawing together main themes.

When CMC is expected to promote instruction, students need to be provided with clear guidelines as to what is expected of them (including the frequency of their participation) and how to participate.

Experience with e-Mail as a Distance Education Medium

Some conferencing systems provide a complex structure to help organize the materials of the conference (or courses). The structures are generally set up using some kind of metaphor of an educational system. These structures often make it difficult for the new users to remember where in the structure they are and to remember how to move among the levels.

A simpler and more generally available alternative to the use of specially-structured conferencing systems is e-mail. We have been using e-mail systems in conjunction with BITNET, to run several CMC conferences and seminars, often linking several internationally located sites. No attempt was made to impose a centralized structure through the system. Indeed, the limitations of the e-mail systems we have been using prevent us from doing this. However the simplicity of use and (for the time being at any rate) cheapness of using e-mail, makes it an attractive alternative to consider.

Our own experience in the use of CMC as a formal methodology of instruction commenced with a sort of computer conference (or rather SEMINAR) which linked up two student groups in similar courses at the universities of Twente (Holland) and Syracuse (in the USA). One of us (Romiszowski) was teaching a course, while on sabbatical at Twente, which was examining the potential impact of new technologies on the practice of education and training. CMC was one of the topics under analysis and so it was natural to wish to get some first-hand experience. At Syracuse University a group of students with similar interests was already using e-mail correspondence to discuss assignments in one of their ongoing courses (even though they were meeting regularly in class session) and were interested in extending experience with the medium.

Further stimulus came from the conference on CMC in Distance Education of the British Open University mentioned above. This experience illustrated (to us, at any rate) that CMC had to be planned as an integrated part of a course, using the strength of CMC as a social communication system for informal contacts between students.

Our interest was in using a simpler system (the existing e-mail linkup through BITNET) but with a certain amount of structure to the interactions that would replicate a number of face-to-face strategies of instruction. We decided to replicate the SEMINAR structure as this would seem to be particularly suitable to the characteristics of the system we would be using. Accordingly, a short (five pages) seminar paper was prepared and a LISTSERVE file was set up on the two universities' e-mail systems, to enable the messages of all participants to be routed automatically to all other participants. A time-frame of one month was set for the discussion of the paper. The structure of the initial paper was carefully planned to focus attention on a relatively narrow band of issues (or so we thought). The chosen theme was an analysis of the potential changes that may occur in the educational technology profession as a result of predicted changes in education/training brought about by the adoption of new communication/information technologies. Half of the paper was devoted to establishing the background assumptions regarding the changes that may take place in education/training systems (e.g., the greater use of distance education methods, remote large databases, expert systems, etc.) and the second half set up a series of predictions for the future of our profession, worded so as to be challenging, thought provoking, and easy to respond to. In practice, though, things did not work out quite as expected.

In all, 47 persons signed onto the network (23 at Twente University and 24 at Syracuse University). Some 100 messages were circulated in a period of some three weeks. However, only 32 of these were directly related to the conference theme. The others were a mixture of test messages, hello messages, chit-chat between individuals, and some technical messages, related to overcoming problems using the e-mail system. The 32 theme-related messages were contributed by only 19 participants, 12 of these sending only one message and five sending two messages. In addition, one participant sent four messages and another sent five, but these were people who had a vested interest and/or organizational function in the seminar.

More surprisingly, the discussion that *did* take place was quite different from the intentions defined in the initial seminar paper. Rather than discussing the predictions made on the effects on our profession of greater use of new technologies in education/training, the participants chose to challenge the basic assumptions that new technologies will in fact have a significant impact! Whereas this is quite a valid and potentially useful discussion topic, it is interesting that the seminar should so easily have got diverted from its (our) original goals. A retrospective study of the dynamics of the seminar suggests that the initial respondents to the paper, who chose to challenge the basic assumptions, then set the tone for the whole experience, in that later respondents tended to reply to the latest messages on the system, rather than to a general framework for the discussion that was set up in a paper some way back in time. It is possible that the chronologically sequential nature of the structure, together with the asynchronous nature of the interaction, combine to encourage the trend that we observed. Be that as it may, several messages input by ourselves, observing the trend away from the original goal and encouraging comment on the predictions for the profession, did not have any effect on the direction of the discussions.

However small the number of messages, the richness of aspects addressed by the participants was quite large. Separate analyses of the structure of the discussion were performed, retrospectively, at both Twente and Syracuse, by different participants. One aspect of interest was that the structure was seen to be different by each of the analysts. One analysis, performed at Twente, identified six main topics of discussion. Another, performed at Syracuse, identified only four main topics and only three of these were present in the Twente analysis. This finding illustrated the somewhat personal view that each participant developed of the seminar. Also, it was significant that the retrospective analysis of the total seminar content did not coincide with the impression that the same participants had formed while participating.

All the analysts agreed that they were surprised by the simplicity of the structure that they uncovered in retrospective analysis. They all had the impression earlier on, during the seminar, that many topics were being discussed and that the structure and dynamics of interaction among participants were much more complex than was in fact the case. This would seem to support the view that the characteristics of CMC create some difficulties for the participant to maintain a clear picture of the discussion over time. While in a face-to-face discussion, several topics may be raised in sequence, their relationship is still kept in memory. The spreading of this discussion over a much longer time period introduces some extra difficulties in keeping an overview/perspective of the discussion as a whole. This effect could well be more pro-

nounced and have more serious consequences in larger and longer CMC sessions.

The observations described above were the basis for the planning of a second CMC seminar, held during the months of April and May 1989. This was planned to examine some of the earlier observed factors somewhat more closely.

Firstly, an attempt was made to generate a seminar topic that would be of intrinsic interest to the target group. Therefore, the topic was generated after a face-to-face brainstorming meeting with Syracuse students. The topic generated was the "Instructional Design" process itself.

Secondly, the seminar paper was kept much shorter than the one we used on the first experience. The object of this paper was only to define with somewhat more precision the theme to be discussed. No specific predictions or questions were included as items to respond to, though a number of possible avenues of discussion was suggested, as an attempt to focus discussion without restricting it as tightly as was the previous intention.

Thirdly, the planned time for this seminar was two to three months and the size of the participating group was to be larger. This would allow us to investigate how the problem of maintaining a coherent view of the seminar's structure changes with the size of the seminar. In order to gather a larger group of participants, an invitation was extended to several other universities. Four universities accepted the invitation and got hooked up to the Syracuse University listserver for the IDCONF (Instructional Design Conference). These four were: University of Twente, Holland; University of Concordia, Montreal, Canada; Indiana University; and the University of Colorado, Denver. In addition, the news spread around and some individuals from other institutions requested to be added to the mailing list.

The overall outcome of this second seminar was, again, not at all as expected. Although a much larger group of potential participants was initially contacted, the number of actual participants who contributed specific content-related messages was smaller than on the previous occasion. Only after several attempts by the seminar moderator at Syracuse to get a discussion theme going, did a significant level of interaction develop. Once more, however, this discussion rapidly drifted away from the expected theme initially defined for the seminar. The initial stimulus messages fed into the system by the moderator were intentionally provocative and focused on the contrast between the prevailing behavioristic approaches to I.D. and the oft-advocated but little practiced cognitive-psychology-based approaches. It was only when he further stressed his point, arguing in later messages that most of these so-called cognitive approaches are in fact behaviorist at heart, that he got some "fish" to bite. Several people commented, some agreeing and some challenging his statement. Before he knew it, we were embroiled in defining behaviorism and cognitive psychology, information processing, and schema theory. This sub-theme came so to dominate the initial discussion, that other messages somewhat more related to the initially selected theme were largely ignored. As a consequence, several early participants dropped out of the seminar discussions, discouraged by the lack of response to their messages.

In time, the seminar ground to a halt, as messages became ever more specialized, ever longer, and ever more abstract. The theorists took over and squeezed out the practitioner and novices. Some of the early participants commented that as the discussion became more specialist oriented, they became less able to contribute any comment that they would think others would be interested in reading. In retrospect, some more vigorous moderation of the process and relegation to a sub-conference of the dominating theory theme may have kept these less confident/less theory-oriented participants "on board." In actuality, the moderation function was intentionally delegated to the students who set up the idea for the conference. They took it the way they wanted to take it and, on the way, lost those who had expected to discuss other aspects of the topic.

In May and June of 1989, during summer school sessions, an attempt was made to revive the IDCONF by imposing more precise, course-related structure to the proceedings. A summer course on aspects of instructional design and development was being taught. The participants of this course were to discuss a series of seminar papers on selected topics. Some of them elected to conduct the discussion of their papers, in part or in some cases totally, on the IDCONF network, rather than normal face-to-face class meetings. The interaction on the CMC system was made part of the expected course activity and was given some weighing in the course grading system. This approach gave much greater focus and purpose to the use of the e-mail system. For one thing, several class attendance sessions were traded for the electronic communication sessions. Students could focus the discussion of their paper to a very precise degree, by concentrating on one topic and by building in stimulus questions/challenges for their colleagues to respond to.

The ensuing debate was a much richer and deeper, as well as a much more focused experience. One spin-off was that several of the "lapsed" IDCONF participants were discovered to still be "lurk-ing" in the system. They were reading the student comments without the benefit of having the original papers to refer to, but the comments were sufficiently stimulating to get these "lurkers" to login and request the papers, so they could also join in the discussion. Some of the papers have been circulated to the IDCONF subscribers, over the e-mail link, and now there is a much larger number of discussants than there were students on the original course. In addition, the depth and detail of this ensuing discussion compares very favorably with that usually obtained in a classroom based seminar session. The course is now over, but the discussion is still continuing. How long it will continue is yet to be seen.

Conclusions: The Problems of Control and Structure

This very brief review of some of the literature on CMC, together with our own initial experiences, leaves us with no doubt that conferencing systems, whether they be purpose-designed or adaptations of existing electronic mailing systems, have much to offer in educational and training settings—not only in distance education but also as a supplementary discussion medium on conventional courses. Experiences such as those of the Open University in the UK illustrate the very important social function that CMC can play in a totally distance education system. They give students a sense of belonging to a community and a medium for communication with other students on their courses (There were even cases of marriages between students, whose first contact had been to discuss assignments in the conferencing systems.) The O.U. experience also illustrated that formal, instructional uses do not develop unless they are intentionally planned into the system of usage and are implemented in a way that generates commitment and develops the necessary skills among the teaching/tutorial force (in case of the O.U. this is a group of several hundred part time tutors, also geographically distributed across the country and with very disparate experience in the use of computers and especially CMC systems—no small challenge).

Our own attempts have been focusing on much smaller projects, which are only components of an educational program that is principally based on-campus using conventional instructional methods. Even so, we have found the task of incorporating CMC into a course, for specific instructional purposes, to be more complex than may appear at first sight. We may focus on two basic problem areas:

(a) **Control:** How do we define the role of CMC in the course as a whole? How do we design a system that is capable of fulfilling that role? Who defines the role? Who (and how) ensures that the ensuing CMC interactions do indeed follow the original plan? What is the appropriate division between system-control and learner-control of the process, the content, the outcomes?

(b) **Structure:** How do we help participants to form, and to keep over time, a clear vision of the structure of the interactions, discussions, and conclusions reached? Should this be the role of a moderator operating centrally on all messages received, or should each participant be encouraged to form a personal view of the structure—if so, then how can we help participants to form a meaningful vision of a multi level/multi speed discussion which stretches over a considerable period of time? etc.

Control

Many proponents of CMC see the principal use and value of the medium as an emancipatory communication medium, totally under control of the users, for whatever purpose they wish to use it. There is no doubt that such a medium of communication is exceptionally valuable in education and was a particularly weak aspect in most conventional distance education systems of the past. The CMC-based common room/coffee-shop is a phenomenon to be welcomed. However, it seems difficult to support the argument that this is the *only* valid use of CMC systems in education. From a general viewpoint, to argue for the restriction of CMC to user-controlled public communication networks, is tantamount to arguing, in the context of conventional educational systems, that the spoken word, or print, should be similarly restricted to unstructured, student-led discussions (i.e., no lectures or textbooks.) However, there seems to be *some* value in this argument, when looking at the experiences in our last conference, where students got the opportunity to bring in their own papers: discussion flared up almost immediately.

The other argument, that CMC only offers significant benefits in its social communication role, deserves more serious consideration. Are other media more appropriate for the communication of structured messages and the control of teacher-designed learning activities? Are they more effective and/or efficient in this role? The evidence, so far, suggests that the asynchronous nature of CMC may in fact offer certain advantages to the conduct of seminar-like educational activities, even when direct face-to-face contact of the discussants is possible. There is evidence of a wider, deeper, and longer lasting involvement of all students. In the last conference this value was sensed by the participating students. They even changed part of the face-to-face sessions into CMC-sessions. However, the "un-moderated" communication between students in both conferences brought about a discussion drifting away from the intended direction. An active role of the moderator in the second conference put people back on track again. The use of CMC as an instructional system has the benefit of the social communication feature of the medium and at the same time, it provides us, when necessary, with moderating, "steering," possibilities for the instructor.

Structure

Structuring CMC is a well-known problem. We are dealing with an electronic medium, which only offers you a very narrow view of the world or the conference: the screen on your PC or terminal. In computer conferencing systems structuring is facilitated by creating a metaphor of the conference; for example, the Campus. As quoted before, it sometimes still is very difficult to know where you are and what is going on. Using electronic mail does not create the problem of not knowing where you are, but one still has the difficulty knowing what is going on. This was especially the case in our first computer conference. In retrospect, people did not even know that the structure was not at all as complicated as they thought it to be. A great part of the structuring could be done by a moderator or tutor in a course.

Computer conferencing systems offer us opportunities to have a pre-fixed structure, we could use in a meaningful way; for example, the division between chit-chat in a special part and content related discussion in another part of the conference system. We think that it is also very interesting and worthwhile for students to do their own structuring. The students' active involvement with the content structure leads to a more meaningful understanding and greater transfer of the knowledge. Therefore, students must be handed some (suggestions for) organizing tools. One could stimulate the labeling of messages in a rational way or download the messages and create pattern notes of the topic areas, which show the interrelationships that may exist between the topics. These structures could be compared to the structures of other students. Another possibility is found in hypertext shells used for organizing the course content as it evolves. The usefulness of this tool should of course be weighed against the additional time commitment that could be necessary for students to learn to use the tool.

Conclusion

In the beginning of this article we expressed our feelings about CMC as a very fascinating instrument for learning. The benefits of extending the educational activities, which are provided by CMC, to the (distance) student, is further reason for seriously considering CMC as a formal part of an instructional system. As we saw and experienced, CMC is not only a highly valuable tool in distance education, but can also have added value for conventional educational institutions. CMC will bring "common" learning and distance education more to each other, which will be profitable for both. There will be a greater instructional richness to fit the needs of all different kind of learners and to deal with the hugh learning and training problems we still have to tackle. ☐

References

Bates, A.W. A mid-way report on the Evaluation of DT200. Paper presented at the Computer Mediated Communication in Distance Education Conference, October 1988 in Milton Keynes, UK.

Davie, L.E. Facilitating Techniques for the Tutor in Computer Mediated Communication Courses. Paper presented at the Computer Mediated Communication in Distance Education Conference, October 1988 in Milton Keynes, UK.

Eldridge, J.R. New Dimensions in Distant Learning. *Training and Development Journal*, 1982, 43-44, 46-47.

Feenberg, A. Computer Conferencing and the Humanities. *Instructional Science*, 1987, 169-186.

Katz, M.M., McSwiney, E., and Stroud, K. Facilitating Collegial Exchange Among Science Teachers: An Experiment in Computer Based Conferencing. Harvard Graduate School of Education, 1987.

Kaye, A. Computer Conferencing for Distant Education. Paper presented at the Symposium on Computer Conferencing and Allied Technologies, University of Guelph, Guelph, Ontario, June 1-4, 1987.

Lauzon, A.C., and Moore, G.A. Fourth Generation Distance Education System: Integrating Computer Assisted Learning and Computer Conferencing. *American Journal of Distance Education*, 1989, *3* (1).

Levinson, P. Connected Education: Progress Report from the front Lines of Higher Learning. *Online Journal of Distance Education and Communication*, May 1989, University of Alaska Southeast, Juneau, Alaska.

Lorentsen, A. PICNIC Project in Computer Networks in Distance Education Curricula. Paper presented at the Computer Mediated Communication in Distance Education conference, October 1988 in Milton Keynes, UK.

Mason, R. The Use of Computer Mediated Communication for Distance Education at the Open University, British Open University, Milton Keynes, UK, 1988.

Quinn, C.N., Mehan, H., Levin, J.A., and Black, S.D., Real Education in Non-real Time: The Use of Electronic Message System for Instruction. *Instructional Science*, 1983, *11*, 313-327.

Romiszowski, A.J., and Jost, K. Lee Computer Conferencing and the Distance Learner: Problems of Structure and Control. Paper presented at the Conference on Distance Education, University of Wisconsin, August 1989.

Rosenberg, M., and Banks, V. The EDTECH Experience. *Online Journal of Distance Education and Communication*, July 1989, University of Alaska Southeast, Juneau, Alaska.

Stern, C.M. Teaching the Distance Learner Using New Technology. *Journal of Educational Technology Systems*, 1986, *15* (4), 407-418.

Toner, P. David. *Computer Conferencing in Formative Evaluation:* The Development and Evaluation of a New Model of Try-out Revision. Doctoral Dissertation, Michigan State University, 1983.

Teleconferencing: An Instructional Tool

Resa Azarmsa

Electronic meetings, convened across a nation or around the world, can be beneficial as well as cost-effective. Every day hundreds of teleconferences are held in different sites, and thousands of interested groups share information, asking questions via telephone line, eliminating the need to travel a long distance or in some instances leave the office.

Soon after the first communication satellite, Telstar 1, launched into a geosynchronous orbit on July 1962, a multi-billion dollar industry started. The capital expenditure for new communication plant and equipment in 1974 was $10.4 billion. This amount increased to $27 billion in 1984. (*Current Business Index*, July 1984). Demonstrably, telecommunication is a growing industry. In 1984, telecommunications ranked 15th in the nation. The teleconferencing industry itself expects total sales (worldwide) of more than $3.7 billion in 1990 (*Teleconference*, Sept./Oct. 1986).

Teleconferencing has been applied to business, medicine, news, military, agriculture, art, and state and federal governments. The business and marketing communities are using video conferencing to meet a broad range of needs. Today teleconferencing has even gone beyond international borders. In Spring 1985, for example, Compaq Computer Corporation decided to hold its annual stockholders meeting and new product announcements simultaneously in Houston, Chicago, Dallas, Los Angeles, San Francisco, New York, and Washington. Beyond those, the conference also was to go to Toronto, London, Paris, and Munich. In all, approximately 3,000 people attended the 90-minute teleconference.

How It All Started

In 1945 Arthur C. Clarke, the science fiction writer, published an article in *Wireless World*, a British publication. Clarke suggested a global communications system using three satellites placed in geosycnchronous orbit at equal distance from each other. Clarke assumed that these three satellites would appear motionless if they orbited the earth at the same speed as the earth's rotation. Because of Clarke's line of work, this idea was not well received in the scientific community.

In the late 1950s, the notion of satellite communications systems motivated a group of scientists at Bell Laboratories to develop a non-geosynchronous satellite communications system called Telstar 1. The first American communication satellite weighed only a few dozen pounds, and was placed in low orbit on July 10, 1962, from Cape Canaveral by a modified Thor-Delta rocket. Because of the low orbit, a massive antenna was needed and continuous communications were impossible. Today satellites are placed in geosynchronous orbit 22,300 miles over the equator. The satellites are at least 180 miles, or 2 to 5 degrees, apart; any closer would cause microwave beams from adjacent satellites to interfere with each other. Satellites orbit at the exact speed that the Earth rotates, so the satellites can always be found at a specific location in the sky. This orbit is called the Clarke Belt in honor of Arthur Clarke.

Types of Teleconferencing

Teleconferencing can occur between two locations (point to point) or between several different locations (point to multipoint). Teleconferencing can use any one of several formats: one-way video/two-way audio, two-way audio/two-way video, picturephone, two-way audio, and computer conferencing.

One-way video/two-way audio is the most common format. Video and audio signals originate from a TV station or a mobile unit, the signals are transmitted to a communications satellite in orbit and then down-linked to an earth station (satellite dish), and the teleconference participants. The participants are encouraged to interact through a direct phone line. The audience can be heard but not seen at the point of origin.

Two-way audio/two-way video is more expensive and mostly used by the television networks. In this format participants are in different locations and can both see and hear one another. The audio and video signals from both sites are transmitted via satellite and received by satellite dish. Picturephone teleconferencing is slightly different from two way audio-video. In this format, the phone line is used to send the video and audio signals. Because of band width limitations, the image is not in full motion. Still images are sent in every 30 to 60 seconds via satellites, microwaves, landlines or other means. The audio, however, is normal. Recent advancement in technology

Resa Azarmsa is a Media Specialist and Adjunct Faculty at California State College, Bakersfield, California.

has improved picturephone teleconferencing quality.

Computer conferencing is at its infancy but it has the potential to be one of the most productivity-enhancing teleconferencing systems. Computer conferencing consists of communication terminals used by participants to access a central computer. The central computer stores text entered by the sender and forwards the information to the recipients when they call the computer. This store and forward feature is the basis for increased productivity (Cowan, 1984). In one respect, computer conferencing is similar to electronic mail. The difference is that instead of thinking in terms of mail boxes, the user should think in terms of file folders. For each participant there would be a file folder. In addition, the file folders can be assigned to topics.

The most inexpensive teleconferencing format is two-way audio teleconferencing. *TeleSpan Newsletter* estimates that 90 percent of the 2.5 million teleconferences in 1983-1984 were audio conferences. (*TeleSpan Newsletter*, June 1983, p. 3). Information is received via phone line and amplified through a speaker phone. The responses are sent from microphone via phone line. This format is effective when individuals know one another and the information to be discussed is not visual.

Transmission Channels

Transmission media (wire cable, fiber optics, laser, etc.) used to carry information from one site to another are classified by different bandwidths. They are narrowband, voiceband, and broadband. The wider the bandwidth, the more information (video, audio, data) transmitted in a given period of time. In addition, there are several transmission schemes. A simplex line permits communication in only one direction. A half-duplex circuit sends and receives information alternately, not at the same time. A full-duplex line allows simultaneous sending and receiving of information. Narrowband has a transmission of 40 to 300 bits per second (bps), or 5 to 30 characters per second (cps). Telegraph and low speed teletypewriters are examples of narrowband transmission. Voiceband transmissions operate from 300 bits to 9,600 bits per second or up to 1,000 cps. Telephone lines are voiceband. Telephone and telegraph could be transmitted through different circuits.

Broadband channels are used for larger information volumes and higher speeds. Broadband operates at the rate from 1,000 to 230,000 bps, or over 100,000 cps. Video signals are transmitted on broadband since they require a wider band because of the nature of the signal and the amount of the information in a television picture. A transmission line called T1 is used to transmit at a rate of 1.544 million bits of information per second. The T1 line can carry videoconferences, phone calls, and data, concurrently. Video transmission is very expensive. However, by compressing the signals, video conferencing is becoming more affordable. A Codec (coder-decoder) device takes the standard, analog TV signal, converts it to a digital data stream, and compresses it by up to 99 percent. The signals are then decoded at the receiving end. This compression enables cost-effective transmission of two-way videoconferencing (Compression Labs, Inc.).

Uplink and Downlink

Videoconference signals usually consist of an uplink from a TV station or mobile transmitter and a downlink to a receiving satellite dish. In an uplink, the signals changes to 6 gigahertz(GHz) through up convertor. Power amplifier amplifies the wattage of signals and leads to the antenna, where the signals are converted to narrow beam for better penetration. Finally, the satellite receiver 22,300 miles away receives the signals. In a downlink, the satellite converts the incoming signals to 4 gigahertz (GHz) and sends the signal to the transmitting antenna. The signals are received by the earth stations. The signals are picked up by the dish and reflected to the feedhorn in the middle of the dish. The feedhorn is in the focal point of the dish. Low noise amplifiers (LNA) boost weak signals and the downconverter converts signals to TV signals (70 MHz). The converted signals pass through a coaxial cable to a satellite receiver. The video signals are processed in the satellite receiver and sent to a TV set for display.

A major advantage of satellite transmission is its declining cost. In 1965 the investment cost was $23,000 per voice channel per year. Today it is in the region of $60 per voice channel per year. Likewise, the costs of earth stations are declining. The first earth station constructed for Telstar cost in excess of $10 million. Now, a powerful earth station receiver costs between $1,000 to $4,000, depending on the size of the receiving dish.

C-Band and Ku-Band

Microwave frequencies begin at about 1,000 Mhz (1 gigahertz) and rise in frequency. Communication satellites operate at microwave frequencies. C-band communication satellites, used by most satellite programmers, operate within the 3.7 to 6 GHz frequency band. Ku-band operates within the range of 11 to 14 GHz. The Federal Communications Commission (FCC) designated the lower portion of the Ku-band (11.7 to 12.2

GHz) as a fixed service band for medium power satellite and the upper half of the Ku-band (12.2 to 12.7 GHz) for high power direct broadcasting service (DBS). The Ku-band is very powerful and does not require a large satellite dish; less than two feet in diameter is sufficient.

A New Solution for an Old Problem

Information is considered an organization's asset. The more information an institution possess, the more accurate and efficient decisions will be made. Computer technology introduced society to the information revolution. Today's computer-powered information revolution is removing burdens of drudgery from the human brain, producing future gains in productivity, and bringing about significant changes in employment, competition, and social changes (Sanders, 1983).

Information accumulates at a phenomenal rate. Despite Arthur Clarke's opinion that the "world suffers from information starvation," the world is obsessed with information. The challenge is the access, distribution, and delivery of information. Remote rural schools, for example, do not benefit from the availability of information as much as do schools in urban areas.

How could diversified programs be offered to information deprived areas? Are students exposed to the latest technology development? Are there enough teachers to teach all subject matters? Are teachers conversant in new technology development? These questions, and many more, often remain unanswered. The resolutions may not be easily attained. Providing comprehensive, cost-effective training to cover both instructor and student in a convenient location seems to be complicated. On the contrary, the new technology offers a solution to an old problem. Distance learning is a new concept that has brought great advantages to education. Aside from the financial problems, there are few limitations to using teleconferencing to enhance the quality of the teaching/learning process. "Telelearning" brings the participants together electronically, eliminating the need to travel to a central location.

A Tool for Learning

Teleconferencing has expanded the horizons of education. Attention is being directed to teleconferencing as a possible way of delivering instructional materials to students in a cost-effective way. The new technology presents information in a concrete fashion and provides students with more accurate verbal and visual facsimiles. The speed and immediacy of information can be comprehended without special effort or skills.

Teletraining has been implemented in learning environment for only a short while. However, current application of teleconferencing already make it possible for educational institutions to include the home or office as learning centers. Many studies have proven the validity of teleconferencing in the teaching/learning process. Some of these studies are examined here:

1. In a study sponsored by the Office of Human Development Services in North Carolina in 1985, the feasibility of incorporating teletraining into a statewide staff development program was examined. The study found that teleconferencing can create a learning environment as desirable as the one created by traditional training, both in terms of amount of learning and attitudes toward training. Of the approaches examined, audioconferencing provided the most cost-effective alternative to traditional training (ERIC Document ED 259 841).

2. A study was conducted of the use of a computer conferencing system developed by Participation Systems Inc. as enhancement for two independent study credit courses in the behavioral sciences, Introduction to Effective Communication and Introductory Psychology. Results of various evaluations of the computer conferencing indicate that the use of teleconferencing courses affects favorably both the quality of the experience and the attitudes and motivational levels of the students. Monitoring of the teleconferencing messages indicated that 90 percent of the students made substantial use of the system. The competition rates for the classes were comparable to the same courses taken in the traditional format, and better than the rates for conventional independent study courses (Haile and Richards, 1984).

3. Pennsylvania State University sponsored a telephone conference network experiment as a cost-effective format for providing inservice training in mental health for individuals who serve the elderly. The study showed that the majority of respondents reported high levels of satisfaction with the telephone conference network system and the specific program in which they participated, and 85 percent reported that they would be able to use the skills learned in the program on the job. The convenience and efficiency of the telephone conference network were the most frequently mentioned strengths of the system (Connell and Smyer, 1984).

4. A project was conducted to revise an existing office occupations curriculum to incorporate the use of telecommunications into simulated office practice and to train teachers in the use of equipment required to implement the updated curriculum. Teachers from two high schools in the school district designed an office simulation

system. A group of high school students were trained to use word-processing software to generate business documents and process them via telecommunications through the various departments of a simulated business. Both the teachers and students who participated in the pilot testing of the curriculum found it to be a valuable learning experience (Burmester, 1984).

5. An instructional program was developed and implemented at Kansas State University to provide career advancement training for respiratory therapy aides and technicians who have had on-the-job training in clinical experience in the basic respiratory therapy skills. The program was implemented under the auspices of TELENET, which is a teleconferencing delivery system based on the campus. The project staff concluded that the use of the TELENET delivery system enabled many individuals to receive training in respiratory therapy who would not otherwise have been able to do so (Tinterow, 1984).

6. Part of the North Slope Borough District, the two schools in the Inupiaq village of Wainwright, Alaska, began to take advantage of computer and audio-conferencing on the subject of using computers as communication tools. They began to use computers to communicate with supervisors in other locations, for math drill and practice, to teach science, and to improve research projects. However, the most interesting use of the computer was as a communications tool for instructional and administrative purposes. Using the computer and one of many electronic networks, students exchanged information with other teachers and with supervisors; administrators sent notes regarding travel schedules, book orders, test scores, and evaluation procedures; and university instructors presented information about the academic consequences of computers (Barnhardt, 1984).

Technological advancement has both facilitated and necessitated the development of teletraining. In developed countries, as Naylor states, distance education is often used to provide traditional education like that usually available in conventional institutions in the Western world. Whereas distance education in the Soviet Union focuses on improving productivity in the workplace, it is used in the United States to provide extension courses, adult basic education, regular postsecondary education programming, and professional continuing education (Naylor, 1985).

Instructional Use

New information technologies will raise the issues of how educational institutions should be restructured; how curricula should be changed and implemented; how teacher/faculty should be edu-

cated to keep pace with a rapidly changing, complex technological society; and what the roles of students and teaching personnel in the learning process should be. Rizzi remarks that institutions must prepare for high technology by having financial resources, expert personnel, equipment, instructional and faculty commitment, and short-range and long-range plans based on well-defined institutional needs, goals, objectives, and resources (Rizzi, 1984).

Applications of teleconferencing in education could be used by students, teachers, and administrators. Teleconferencing applications for students may be called instructional, and when used by teachers and administrators it is inservice training or professional development. Together they help students learn:

- Teaching subjects (math, science, computer, etc.) in remote areas where experts are not available.
- Teaching new technology where access to the information is limited and costly.
- Teaching a subject in a big geographical area where standard instruction is crucial.
- Teaching materials in the multi sites where the participants' interaction is desired.
- Teaching materials for great numbers of students where demonstration is critical.
- Teaching where a well prepared presentation is necessary.
- Teaching a foreign language in multi-sites where a phone line is available.
- Teaching subject matters on different campuses where simultaneous instruction is desired.
- Teaching subject matter for adult education programs where attendance in class is a problem.
- Seeking experts' opinions on specific educational and social trends and issues when the experts are not physically nearby.
- Watching special academic events, such as a speech, debate, or a spelling contest.
- Listening to keynote speakers who can not travel extensively and interaction is required.

For inservice training or professional development:

- Routine inservice training.
- Updating information.
- Immediacy of information.
- Timeliness of information.
- Inexpensive training for a large group.
- Renewing credentials.
- Multi-connection between different sites for interaction.
- Flexible hours, after school.

- Watching different models of teaching at different sites.
- Electronic visitation of facilities.
- Electronic class reunion.

Many other applications can be added to the list but one very important point must be remembered: teleconferencing is not intended to replace the classroom teacher but, rather, to extend the classroom beyond its immediate walls. Also, in applying this technology to instruction, caution should be employed because teleconferencing lacks the complexity of face-to-face instruction and interaction. In addition, the sophisticated level of equipment, in some cases, might be intimidating.

Summary

Teleconferencing has great potential for increasing the effectiveness of instruction. The telecommunications revolution offers different channels of instructional communications, from telephone, microwave, and satellite to computer. Teleconferencing can be cost-effective and time-saving. In addition, teleconferencing increases instructional productivity and makes teachers' inservice programs more efficient and available. □

References

Barnhardt, C. *Let Your Fingers Do the Talking: Computer Communication in an Alaskan Rural School*, 1984. (ERIC Document Reproduction Service No. ED 241 242.)

Bjorklund, J., and Fredmeyer, J. Keeping Current Via Teleconferencing. *Journal of Extension*, Summer 1985, 21-24.

Burmester, A. *Telecommunications. Final Report*, November 1984. ERIC Document Reproduction Service No. ED 255 679.)

Connell, C., and Smyer, M.A. *Training in Mental Health: Evaluation of the Telephone Conference Network*, November 1984. (ERIC Document Reproduction Service No. ED 255 780.)

Cowan, R. *Teleconferencing Maximizing Human Potential.* Reston, VA: Reston Publishing Co., 1984.

Cowan, R. *Current Business Index*, July 1984.

Eastman, S.T. Teleconferencing in Education and Business. *Feedback*, Summer 1985, 8-11.

Easton, A.T. *The Satellite TV Handbook.* Indianapolis: Howard W. Sams, 1983.

Haile, P.J., and Richards, A.J. *Supporting the Distance Learner with Computer Teleconferencing*, October 1984. (ERIC Document Reproduction Service No. ED 256 293.)

Johanson, R. *Teleconferencing and Beyond.* New York: McGraw Hill, 1984.

Kurland, N.D. Have Computer, Will Not Travel: Meeting Electronically. *Phi Delta Kappan*, October 1983, 122+.

Nayler, M. *Distance Education. Overview. ERIC Digest No. 44*, 1985. (ERIC Document Reproduction Service No. ED 259 214.)

Nayler, M. Office of Human Development Services. *Expanding Human Services Training Through Telecommunications: A Day Care—Head Start Study. Executive Summary*, January 1985. (ERIC Document Reproduction Service No. ED 259 841.)

Olson, D. Satellite Session for Manjor Meetings. *E-ITV Magazine*, November 1985, 30-31.

Portway, P. What Teleconferencing Adds, Not Eliminates. *Office*, April 1984, 101+.

Rizzi, R.A. *Impact and Implications of Technological Change on Educational Institutions*, March 1984. (ERIC Document Reproduction Service No. ED 249 944.)

Ratcliff, J.L. Statewide Teleconference Educational Programming: Strategic Planning Issues. *Catalyst*, 1985, *14*(1), 12-16.

Rogan, R.G., and Simmons, G.A. Teleconferencing. *Journal of Extension*, September/October, 1984, 27-31.

Sanders, D. *Computers Today*. New York: McGraw Hill, 1983.

Sanders, D. Telecom VI: World's Largest Teleconferencing Show. *Teleconference*, September/October 1986, 10-11.

Thiel, C.T. Teleconferencing: A New Medium for Your Message. *InfoSystems*, April 1984, 64-68.

Thiel, C.T. Turn on to Ku-Band. *Satellite Orbit*, December 1985, 38+.

Tyson, J. Satellite Sessions for Decision-Making. *E-ITV Magazine*, November 1985, 27-29.

Tyson, J. *US Bureau of Economic Analysis*, 1984.

Tyson, J. *Videoconference Primer*. Compression Labs, Inc., San Jose, CA, 1985.

Winn, B. *et al.* The Design and Application of a Distance Education System Using Teleconferencing and Computer Graphics. *Educational Technology*, January 1986, 19-23.

Computer Conferencing: Models and Proposals

Valarie Meliotes Arms

I have participated in two computer conferences which function quite differently from each other.* These conferences have proven to be a powerful tool for humanists to share knowledge and to collaborate on writing and research projects by allowing scholars from across the country, in some instances, internationally as well, to communicate with one another. I have also edited a special issue of the IEEE *Transactions on Professional Communication* (vol. PC-29, no. 1, March 1986) on computer conferencing. All the articles cited here are from this issue, which evidences the importance of computer conferencing in varied disciplines and settings. As an historical note on computer conferencing, the most important early work in the humanities was initiated by the Western Behavioral Sciences Institute in La Jolla, California. IBM has made computer conferencing available free to scientists throughout Europe and Israel, connecting them with American networks. Courses in computer conferencing are now available from the NY Institute of Technology, The New School for Social Research, and the University of Colorado. Humanists are beginning to realize the potential as much as the scientists have.

Marshal McLuhan offered some interesting ideas that parallel what we do in computer conferencing, Paul Levenson, in an article entitled "Marshal McLuhan and Computer Conferencing," says that McLuhan's work "often took the form of numerous free-standing commentaries, usually not more than a few pages, each self-sustaining yet revolving around some sort of central theme. This holographic style turnes out to have much in common with the commentaries produced by participants in a computer conference. The electronic group

*The host computers are at the NJ Institute of Technology for the FIPSE group and at the New York Institute of Technology for The Fifth C.

Valarie Meliotes Arms leads the technical writing staff in the Department of Humanities and Communications, Drexel University, Philadelphia, Pennsylvania.

production of dialogue fulfills McLuhan's oft-stated desire to overcome the boundaries of static print on a page and shows that literacy may actually be promoted rather than eclipsed by the electronic revolution." (IEEE PC March, 1986)

What We Need

To begin our computer conferencing, then, let's discuss what we actually need: a phone and a computer, obviously at one point, and at the destination, another phone and computer for somebody to pick up the messages. If you think in terms of your post office box, you send your message through your computer and phone to a post office box. When I'm ready to pick up the message, I go to the post office box, via my phone and computer, and can receive it at my convenience. I can decide to use either a prepared message to respond to you or to compose my message on-line. The composition on line tends to be more informal as opposed to a structured paper which I might have thought about some time and revised before I uploaded it onto the system. The receiver of the message can choose to respond or not to respond, which can be very disconcerting because the sender cannot tell if the lack of a response implies indifference, disagreement, or mechanical difficulty.

The host computer, which can be housed in a place like NJ Institute of Technology or NY Institute of Technology, holds the message until we decide what to do about it. To turn to the two groups that I mentioned, let's begin first with The Fifth C, which is Computers for the Conference on College Composition and Communication. The Fifth C group meets informally on line, which is in the spirit of the group's inception.

A Special Interest Group

Several people—Hugh Burns, Helen Schwartz, Kate Keiffer, and Charles Smith, among others, including myself—met several years ago when 4 C's was in San Francisco. We decided that there was a need for a special interest group on computers and composition. Bruce Appleby thought of the name, "The Fifth C," and we have Michael Spitzer to thank for a grant which made possible putting The Fifth C on line. We function as resource for one another. For example, we share writing exercises, ideas on research projects, and critiques of each other's articles. When one member of the group had a problem with an article that she had written and needed to respond to a letter, she asked for advice on how to respond to her critic. Dawn Rodriguez has been known to describe the group talking together as an invention tool. We exchange ideas, test out, and, in a sense, pre-write to one another.

Bryan Pfaffenberger has noted in an article "Research Networks, Scientific Communication, and the Personal Computer," that the personal computer has made it especially easy to give young researchers access to the "network of established researchers and to contribute their ideas in ways that abolish discrimination. Personal computers in computer conferencing obliterate social barriers and status; computer conferences rectify the network's tendencies toward elitism and internal stratification. Research shows that they foster a strong sense of group identity. Studies on how scientists exchange information note that they often make research results more explicit in informal communications than they do in later journal publications. Existing networks are closed to young minorities, geographically isolated people; it is, in a sense then, that computer conferencing can truly be considered an egalitarian way of communicating." (IEEE PC March, 1986)

Informality

Another writer who has commented on the nature of such coffee-clatch communication describes in his article, "Chit Chat to Electronic Journals: Computer Conferencing Support Scientific Communication," how the British library research and development department decided to use an electronic journal (IEEE PC March, 1986). As David Pullinger goes on to explain, professional researchers, both industrial and academic, communicate in all sorts of different ways, by casually meeting over coffee, by exchanging drafts of articles, and by meeting at seminars, colloquia, and conferences, such as this one. "Indeed, the very thing that the journal does not tend to do is to inform other researchers in the core group of new material. It has been imparted before that stage by some invisible, but usually clearly perceived, network." Thus, in a very valuable sense, the computer conference is like a coffee-club which allows us, in a democratic way, to brainstorm with one another.

The FTSG group, which is the FIPSE Technology Study Group, works in a different fashion. FIPSE wanted to find a way to disseminate the ideas that many of its project directors had on using technology. They brought together, on line, the FTSG Group. We began very formally by having a meeting; the people who attended had their photographs taken; the photograph page was distributed to all of the members of the group because these were not people who even moved within the same domain. The diverse group included educators from biomedical sciences, engineering, architecture, and English. These people needed to understand one another's discipline before they could begin to talk about what they shared in common. To facilitate bringing ideas together in such a diverse group, there were formal meetings.

Andrew Feenberg, who has written an article "Network Design: An Operating Manual for Computer Conferencing," was at our first meeting to explain the role of the moderator (IEEE PC March, 1986). A moderator is much like the discussion leader in a regular conference. He noted that the moderator could be a facilitator who helps to decide procedural questions, such as how to introduce a new topic, when to close a topic, and how to reach a decision in the length of comments; or he could be an intellectual leader, deciding what information to transmit, what to do about it, setting the context, suggesting readings, and setting norms for acknowledging collaboration. He noted that moderators of both types need to recognize contributions (if nobody recognizes a comment or responds to it, you tend to feel hurt) and need also to weave comments and conferences together. Elaine Kerr, in an article, "Electronic Leadership: A Guide to Moderating On-Line Conferences," draws on her experience for several years on the EIES Network, and notes that the social context is important in a face-to-face meeting, that strong leadership has been shown to be required in this medium if groups are to be successful (IEEE PC March, 1986). The absence of pressure to sign on-line and to participate creates the need for strong and active leadership. Elaine was part of the FTSG Group in a subsequent meeting where she discussed further the techniques used on-line.

In this highly structured FTSG Group, where we have a given topic to report on in order to write chapters in a national report, which we hope to publish, there are assigned tasks with deadlines that each one of us must meet. The responsibility here is in contrast to the coffee-club, where we sign on only when we feel a need. The group is funded and has a specific deadline. One of the ways that we help communicate to one another is by mailing information as well as using the computer conference; the papers from each of us, describing our individual projects or other ideas that we might have, are mailed out through FIPSE so that we have reading material to go over before we even sign on-line in many instances. Whichever model would be appropriate for your group, you would have to contend with individual styles of response.

Interpretation Difficulties

While some writers may respond without reflection to a message, as if it were a phone call, others will want to plan their comments. Yet, in each case there is no body language, no visual clue, no facial expression, no eye contact, no inflection in our voice to help the reader interpret what it is we're saying, nor can we see the puzzlement in the receiver's face, or the expression of distain that they may have when they read our message. Without these visual clues there is a great deal of misinterpretation of comments on a computer conference.

Sara Kiesler of Carnegie-Mellon has been studying participants in computer conferencing as a behavioral scientist. It seems that when people "talk" on computers they tend to be more rude, to use profanity, to have emotional outbursts, which we call flaming, and that people in computer conferencing take longer to agree, that their final decisions tend to be more extreme. When they finally do agree, they tend to be very rigid about change. Some of the other behavior which she has noted has brought forth its own jargon to accompany computer conferencing. Some of the behavior, such as "lurking," can be very funny when you consider that somebody is almost hiding out in a trench coat peeking around corners and reading the messages, but it can also be very annoying because it brings up the question of people reading and perhaps taking the information from the computer conferencing without the author's approval; and so it leads to some of the problems that we need to consider in setting up a computer conference. Flaming, which is emotional outburst, likewise causes problems in computer conferencing because we don't know, again, what is appropriate behavior in this context. Bad spelling is another problem. Do we set up rules of grammar? Does spelling count or not? And if we say that it doesn't count, does it still influence the way a message is perceived or the authority of the speaker?

Given all of these problems, we might decide then when we set up our own computer conference how to go about establishing a protocol. As in any good writing task, it is necessary to consider the audience. Imagining the group to be addressed, we need to say what is the purpose of the computer conferencing; who is the audience; why are they using the system; what are their common goals; what is their access to hardware and software; how comfortable are they with that? We might ask if they own their own equipment or if it will be necessary to provide them with equipment. Having considered some of these basic questions, we might think of some general principles for establishing the protocol.

Twelve Protocol Questions

For the imagined person in the computer, protocol ought to consider twelve important questions. (1) Should there be a face-to-face meeting to initiate and perhaps annually renew contacts? (2) Should participants sign a contract stating how often they will log on and what their responsibility is to contribute? (3) How will quotes or ownership of ideas be designated and safeguarded? (4) How important are the English standards: spelling and grammar? (5) What will be the moderator's role; how will the moderator be chosen, what authority does the moderator have? (6) What budget does the moderator have for incidentals like phoning people to prod them into getting on-line? (7) Mailings—will it be necessary to send copies of other articles or even of the transcript of the computer conference itself, for those who have not been able to be on-line for one reason or another? (8) What help can be provided; whether the help is on-line, by phone, or adequate documentation of a system? (9) What are the specific tasks that face the group? Is it an exchange of ideas, the production of a paper or book? (10) What separate metaconference needs to be established for discussing the experience of conferencing itself? (11) Indexing—who will be responsible for defining key words and making it possible to find entries within the system? (12) What social norms will be acceptable; are there any lurkers who should be tolerated? Having considered these protocols it would be possible to set up a computer conference based on the group's particular needs, which may turn out to lead to a very fruitful collaboration for humanists. Writing groups have been used successfully for years; conferencing simply makes more varied groups possible. ☐

Learning Over the Lines: Audio-Graphic Teleconferencing Comes of Age

Michael K. Gardner, Sidney Rudolph,
and Gabriel Della-Piana

Until recently, rural America had been slowly losing the battle to provide its students with a quality educational experience similar to that found in urban and suburban areas. The reasons were simple enough. Rural schools were small in terms of faculty size and tax base. They were in isolated areas that made it difficult to attract top young teaching talent. The results were overworked faculties teaching outside their areas of expertise, and students who found their first year of post-secondary education a little like a ride on the roller coaster at Coney Island: something that has a lot of ups and downs, and requires hanging on for dear life.

The problem of how to improve rural education seemed unsolvable, given economic reality, until the advent of inexpensive microcomputer technology during the 1970s and 1980s. This technology is currently being applied to the problem, and remarkable progress is being made. In this article we will describe one such application of computer technology to the problems of rural education, and an evaluation of the success of that approach. The approach is called audio-graphic teleconferencing, and it was used at a number of sites in Utah to provide advanced curricular offerings to schools that would otherwise not have offered these courses. We feel that this approach holds a great deal of promise for narrowing the educational gap between rural and non-rural schools.

Audio-Graphic Teleconferencing

Audio-graphic teleconferencing, as we have developed it, consists of two types of locations or "sites." The site where the instructor is located, and from which classes emanate, is called the "teaching site." The sites where students receive instruction are called "remote sites." The sites are interconnected so that two-way audio (voice) and computer graphics (data) transmission is possible between all sites. The connections are made over standard telephone lines, so that no expensive dedicated wiring or radio or microwave transmission system is necessary (although such systems could certainly be used if they already existed). The use of existing communications systems, such as telephone lines, keeps the cost of audio-graphic teleconferencing within the budgets of most school districts.

At each site certain equipment is necessary. For audio communication at both the teaching and remote sites, a telephone, a microphone, and a conferencing device (e.g., a convener) containing an interface to the telephone and microphone were used. For graphics communication at both sites, the following equipment was used: (a) a personal computer with 256 kilobytes of random access memory, two double sided, double density disk drives, a 1200 bit-per-second internal modem, a high resolution color graphics adapter, and a serial adapter for a digitizing tablet; (b) a color television monitor; (c) a digitizing tablet, electronic stylus, and power supply; (d) a power strip; and (e) a computer software package that allowed communications, "blackboard" operations on the graphics system, and authoring of graphics to be used for upcoming lessons.

Class would begin by the instructor "dialing up" the remote sites to establish voice and graphics links. Students at the remote sites were seated near the front of the classroom so that they could easily see one of the two television monitors that conveyed the graphics information. A classroom management person at each of the remote sites would take the attendance role, and instruction would begin.

Instruction followed a lecture-type presentation format, with the instructor speaking over the audio system and using the graphics system as an electronic blackboard. Material could be presented over the graphics system in one of two ways: the instructor could write or draw material by hand on the digitizing tablet and have it appear on the television monitor, or previously prepared graphics material (e.g., a physics problem) could be recalled from a disk file and presented in total. Students could interact with the instructor either by asking questions or responding to the instructor's queries. Because communication was two-way, a question would interrupt the instructor's transmission. In addition, a student could solve a problem at the

Michael K. Gardner is Assistant Professor, Department of Educational Psychology, University of Utah, Salt Lake City. Sidney Rudolph is Research Associate Professor, Department of Physics, University of Utah. Gabriel Della-Piana is Professor, Department of Educational Psychology, University of Utah.

"board" by using a digitizing tablet to work a problem presented by the instructor. The results of the student's effort would be transmitted simultaneously to all of the other sites, allowing both the instructor and other students to view this student's work.

Assignments and tests were sent from the instructor to the various classroom management individuals via school courier or the mail. When students had completed the assignments or tests, the management personnel would return them in the same way to the instructor. The instructor would then grade the assignments and return them to the remote sites, again using a school courier or the mail. This process was not as efficient as we would have liked, and we contemplate using an electronic transfer process (e.g., sending computer files via modems) in the future.

The key attributes of audio-graphic teleconferencing are that it is: (a) fully two-way in both audio and graphic transmission, (b) it is less expensive than most other forms of two-way communication, such as two-way television, and (c) it allows an expert instructor at one location to service small groups of students simultaneously at several remote locations. These attributes make it possible for small rural schools to band together, cooperatively providing more high quality classes than any one school could provide by itself. For instance, school A may have a certified chemistry teacher on its staff but no certified physics teacher. School B may have a certified physics teacher but no certified chemistry teacher. By using audio-graphic teleconferencing to share resources, the two schools can benefit from each other's strengths. In addition to sharing between schools, regional education centers and universities can lend some of their expertise to help supplement offerings at small, rural schools.

Evaluating the System at Work

When we began to develop audio-graphic teleconferencing several years ago, we faced the criticism that teaching over speakers and computers just isn't as good has having a teacher present in the classroom. This criticism has a good deal of face validity; however, we wished to test the claim directly since experience has demonstrated that a good deal of what we think is true, upon closer investigation, isn't.

A Quasi-Experimental Study

Two of us performed a quasi-experimental study comparing the performance of students taught by audio-graphic teleconferencing with those taught by traditional "live-teacher-in-the-classroom" instruction (for a full report, see Rudolph and

Gardner, 1986-87). In this study, conducted at Bountiful High School, Bountiful, Utah, students ($N = 11$) enrolled in physics during one meeting period received 11 weeks of physics instruction using audio-graphic teleconferencing. Students ($N = 17$) enrolled in physics during a second meeting period received the same 11 weeks of physics instruction (that is, the same topics were covered) from the same instructors using the traditional classroom teaching approach. Students' performance on both objective and computational tests on the material covered during the intervention showed no significant differences between the two methods of instruction. In other words, there was no evidence that traditional teaching was superior to audio-graphic teleconferencing.

Two Evaluations of Teacher and Student Interest in Audio-Graphic Teleconferencing

This somewhat counter-intuitive result led us to study audio-graphic teleconferencing as it was being used in larger settings within the state of Utah. We performed two program evaluations (Gardner and Della-Piana, 1986a, 1986b) of audio-graphic teleconferencing systems being used in two different geographic and demographic areas. Because it was not possible to perform quasi-experimental studies in these schools, we concentrated on assessing teacher and student interest in the technical system and reactions to it. We will discuss each of these in turn, paying special attention to the commonalities that arose across the evaluations.

Audio-graphic teleconferencing in Davis County, Utah. Davis County, Utah, is a primarily suburban county north of Salt Lake City. Davis County had implemented audio-graphic teleconferencing in three high schools to provide high-end courses, such as "Advanced Placement" (AP) courses, whose enrollments in any single school would have been too small to justify their being taught. During the year we evaluated the Davis County audio-graphic teleconferencing project, one course—AP physics—was being offered at the three high schools. The instructor for the course was a regular teacher at one of the high schools with a good background in physics.

We observed the system in action, had students at each of three high schools complete a short questionnaire, and interviewed the instructor and a sample of students concerning their experiences with audio-graphic teleconferencing. We found that student acceptance of the teleconferencing was fairly good. When asked, "How much have you enjoyed the computer assisted AP physics course?" students gave an average response of 5.4 on a scale of one to seven, where one denoted "Not at all" and seven denoted "A lot." Also, when asked, "Was the method of instruction interesting?" the mean response was 3.0 on a scale where one stood for "Very interesting" and seven stood for "Not at all interesting." Students seemed particularly pleased with the computer graphics presentation of physics problems. We asked them, "Were the problems presented on the computer screen by the teacher helpful in understanding the concepts being presented?" Their average response was 2.8, where one represented "Very helpful" and seven represented "Not at all helpful." A final question of interest was, "How hard was it to learn from the teacher and the computer, compared to a regular class with a teacher in front of the class?" The average student response to this question was 3.2, where one indicated "Much harder than a regular class" and seven indicated "Much easier than a regular class."

Our findings indicated that students enjoyed the class and its attendant computer graphics, but still found it a more difficult learning environment than a traditional class. Interviews clarified the nature of some of the difficulties that students experienced. Some students noted that system electromechanical failures caused valuable classroom time to be wasted. Further investigation showed this was primarily due to the failure of duplex phone line jacks. Once the problem was diagnosed, a supply of extra jacks was purchased, and system failures decreased.

Another problem students mentioned was the great deal of time it took to get homework and tests graded and returned. The problem stemmed from the use of a school courier or the mail to transfer papers. Although the schools in Davis County were not separated by great physical distances, this transfer process could extend the amount of time it took to turn papers around to over one week. Clearly such a delay in feedback to students is unacceptable, and we plan to use an "electronic mail" system to transfer work in the future. By transferring files via modems, the turn-around time should be cut to two to three days.

Finally, a major problem at one of the schools involved the lack of adequate classroom supervision. In this school unsupervised students attempted, often successfully, to sabotage the teleconferencing system by playing music over the audio link or by overwriting material on the graphics screen. This disrupted both the problem school and the other schools which were electronically linked up with the problem school. While it might be possible for us to safeguard the system in such a way that it would be difficult or impossible for students to sabotage it, we think these incidents point up a more basic issue. Audio-

graphic teleconferencing can help improve the quality of instruction in schools, but it does not relieve administrators from the responsibility of supervising classes. It is *imperative* that each teleconference class have a classroom management person present to pass out assignments, make small adjustments in the technical system, and maintain an atmosphere conducive to learning. Such a person does not need to be a teacher, but it does need to be someone able to maintain classroom order. Our experience with the one problem school we have described convinces us that all of the benefits of audio-graphic teleconferencing can be wiped out if classroom management is ignored.

Audio-graphic teleconferencing in the Northeastern Utah Educational Services area. Northeastern Utah Educational Services is a regional service organization that assists school districts in rural northeastern Utah. It had implemented audio-graphic teleconferencing to teach three courses—physics, AP history, and college level English composition—to three very small, rural schools. The schools were in farming communities, and were isolated from one another as well as from urban centers. The history teacher was a regular faculty member at one of the three schools; the physics and English teachers were faculty members at an area vocational education center not directly affiliated with any of the three schools.

We observed the system in action, had students fill out a brief questionnaire, and interviewed the instructors and a sample of students, just as we had in Davis County. Once again we found acceptance of audio-graphic teleconferencing was reasonably good. When asked, "How much have you enjoyed this computer-assisted course?" students had an average response of 4.6 on a seven point scale where one denoted "Not at all" and seven denoted "A lot." In addition, we asked, "Was the method of instruction interesting?" The mean response was 3.5, where one indicated "Very interesting" and seven indicated "Not at all interesting."

As in Davis County, students found the graphic presentation of problems to be a strong point of the system. When asked, "Was the material presented on the computer screen by the teacher helpful in understanding the concepts being presented?" students gave an average response of 3.1, where one represented "Very helpful" and seven represented "Not at all helpful."

Students in these rural schools also saw the audio-graphic teleconferencing system as a slightly more difficult learning environment than a traditional class. We asked them, "How hard was it to learn from the teacher and the computer, compared to a regular class with a teacher in front of the class?" The average response was 3.7, where one stood for "Much harder than a regular class" and seven stood for "Much easier than a regular class."

We asked students in the Northeastern Utah Educational Services area two questions that we had not posed to Davis County students. First we asked, "Which is a better way to take a class: as a computer-assisted class with other schools, or the traditional way with a teacher in front of the class and no computer or other schools?" Students split on this question, giving a mean response of 4.0, where one represented "A traditional class is a lot better" and seven represented "A computer-assisted class is a lot better." Next we asked, "If you had the choice between taking a computer-assisted course on a subject not normally taught at your high school or waiting until college to take the course, which would you choose?" Students overwhelmingly preferred the computer-assisted option, with an average response of 2.4, where one indicated "I'd take the computer-assisted class" and seven indicated "I'd take the course in college." Thus, while there may be problems with the audio-graphic teleconferencing system in its current incarnation, it is seen as a viable option when regular staff does not exist to meet important course needs.

Interviews revealed many of the same problems occurred in these rural schools that had occurred in the suburban schools discussed earlier. One new concern arose, however. Different teachers had used the computer's graphics capabilities to differing degrees. This was partly due to the fact that some subject matters, such as physics, were more readily suited to graphic presentation than were others, such as history. But students seemed to think that courses that did not make much use of graphics could have benefitted from its innovative use. Upon reflection, we agree with the students. A course such as history could have made use of graphics through devices such as maps showing troop movements during wars and illustrations of famous historical figures. Other uses obviously exist; only brainstorming is necessary to discover them. Our questionnaires lead us to believe that students will respond positively to the increased use of graphics in all audio-graphic teleconferencing courses.

Conclusions

We may have been too optimistic when we initially concluded that audio-graphic teleconferencing was not demonstrably worse than traditional classroom instruction. Students have told us they find teleconferencing to be a more difficult learning environment than having a live teacher in the classroom. But we must keep in mind what the real question is. It is not, "Should we have live, qualified teachers or audio-graphic teleconferencing?" Instead, the question is, "Should we have audio-graphic teleconferencing or classes taught by teachers not expert in a subject—or possibly no classes at all in certain subjects?" To this question our students' response is clear. Audio-graphic teleconferencing is clearly a better alternative than waiting until college.

To those involved in the development of audio-graphic teleconferencing systems, as we are, we would give the following advice:

1. Make full use of computer graphics wherever possible. The one thing audio-graphic teleconferencing can do that a traditional teacher with a blackboard cannot is to provide exciting, attention-grabbing visual presentations. This benefit should be exploited to its fullest.

2. Make certain that adequate classroom management is present in all remote sites. Local administrators need to be reminded that just because instruction is coming from elsewhere does not mean that they can assume a class will run totally on "autopilot." They are responsible for keeping order in their classrooms.

3. Automate the transfer of material between sites so that students can receive feedback in a reasonable period of time. Our experience is that school courier or regular mail is too slow. Electronic mail is a viable alternative.

4. Continue trouble-shooting an audio-graphic teleconferencing system for at least several months after it has been installed. We discovered a large percentage of recurring system problems were the result of one or two easily remedied electromechanical problems. After these were fixed, system failure was infrequent.

We think the future for audio-graphic teleconferencing looks bright. We feel that it offers the potential to enrich rural curricula by increasing course offerings. This can be accomplished through the use of technology and the sharing of educational resources among the various agencies charged with serving a geographical area. Although the courses discussed here were academically oriented (AP courses), audio-graphic teleconferencing could equally well be used with other content domains, such as vocational education. The technology is general, and the applications described here are only examples. School districts and state offices of education must now make the tough dollars-and-cents decision to invest in the technology, if they wish to reap the benefits. □

Use of Audiographic Technology in Distance Training of Practicing Teachers

Dennis R. Knapczyk

Research conducted by Helge (1979, 1983, and 1984), Lehr and Harris (1988), Slavin, Karweit, and Madden (1989), Treadway (1984), and others has shown that instructional and administrative staff of many rural schools are ill-prepared for the responsibilities they encounter in providing educational services to students. Giving better preparation to school personnel will require teacher training institutions to change the manner in which they structure and deliver training experiences, especially those offered to personnel already working in rural communities (O'Connor and Rotatori, 1987, Pelton, 1983).

Providing off-campus coursework to teachers in rural schools presents major challenges for training institutions. Models of distance education that use correspondence courses, on-site instruction requiring travel by professors, interactive television, teleconferencing, and other approaches have produced an array of options for planning and delivering programs (Anderson, 1989; Condon, Zimmerman, and Beane, 1989; Egan, McLeary, Sebastian, and Lacy, 1988; Keegan, 1983).

Audiographic technology provides an additional option for linking universities with rural schools and assisting them in devising new models of personnel preparation (McConnell, 1983; Williamson, 1983). Audiographic technology can aid in establishing partnerships between universities and school corporations and support such activities as identifying the needs of personnel, utilizing local resources in training activities, and devising on-the-job training experiences suited to the demands of the school environment (Knapczyk, 1989). This article describes the use of microcomputers and audiographic technology in providing off-campus

Dennis R. Knapczyk is with the Department of Curriculum and Teacher Education, Smith Research Center, Indiana University, Bloomington, Indiana.

graduate level training to teachers of at-risk and mildly handicapped students in several rural communities of southern Indiana.

Overview of Audiographic Technology

Audiographics is a hybrid technology that permits simultaneous transmission of voice communication and graphic images across local telephone lines. Audiographics is interactive and provides users with the capability of speaking with one another, sharing text and graphic images, and annotating images displayed on monitors. Graphics can be stored, sequenced, and used as part of large-group presentations, small-group conferences, and individual reporting and feedback sessions. Multiple user sites can be networked together in a conference call format with each site having full interactive capabilities.

The audiographic system used in the teacher training program at Indiana University is one developed by American Telephone and Telegraph (AT&T). The system contains the following components:

a. Quorum Conference Phones.
b. Work Group (WGS) personal computers with VGA boards and high resolution monitors. The computers have hard disk drives and 640KB of memory.
c. Overview Scanners that are desktop devices that capture half- or full-page images. Almost any type of image can be scanned with high resolution, e.g., photographs, charts, hand written documents.
d. SCANWARE communication software package that provides the ability to scan and capture images, transmit images to other locations, display and annotate images on the monitors, and sequence and store multiple images in the form of slide presentations.
e. Bridge Modems that operate at 4800 bps and give the capability of both point-to-point and multi-point communication networking.
f. Printers.

Additional components, such as external microphones, large-screen monitors, graphics tablets, and audience response devices, are being incorporated into the system to enhance its capacity for training. In addition, audiographics is being used in conjunction with less costly telecommunication systems, such as electronic mail and bulletin board networks, to provide a full range of options for large-group, small-group, and individual teletraining activities.

The audiographic system utilized by Indiana University operates across two sets of dedicated telephone lines. One line is used for voice transmission and the other links computer terminals.

Overview of Indiana University's Program

The At-Risk Program at Indiana University is field-based and offers training to practicing teachers in rural communities who otherwise would be unable to participate in university coursework. The program recruits instructional personnel from selected schools within a region so that training experiences can be adapted to the characteristics of the schools in which the trainees teach. Teams of two to six teachers from schools proximate to one another form cohort groups for the program. About 30 school personnel representing seven schools in two regions of southern Indiana are currently enrolled in coursework.

The program provides twelve credit hours of graduate coursework: six credits of academic instruction and six credits of supervised practicum. Coursework is designed to improve the expertise of school personnel who work with at-risk and mildly handicapped students by emphasizing educational practices and methodologies that promote self-reliance and successful performance in regular education settings. The program also addresses restructuring of both classroom and schoolwide practices and fosters collaboration and collegiality among personnel. Training activities are developed and delivered in partnership with participating school corporations. Local personnel assist in planning training experiences, team-teach in the academic courses, and take an active part in program evaluation.

The sequence of training experiences is delivered in off-campus regional sites using a distance education model. A two-way communication network is established between Indiana University and the schools. The network uses audiographic technology to assist presentation of coursework, enhance supervision of practicum activities, offer counseling and advisement, and connect trainees with library services and other campus resources. In addition, microcomputers in each of the schools are networked with one another through the university mainframe computer to give trainees access to electronic mail and bulletin board services. Networking schools within a training region allows teachers from different schools to collaborate on training activities, share progress on projects, and provide suggestions for alternate teaching practices.

Use of Audiographics in Delivering Academic Courses

Academic courses cover methods for assessing and teaching at-risk and mildly handicapped students. They are delivered in off-campus regional sites that are proximate to the schools in which trainees teach. Teachers enrolled in the courses commute to the regional sites and participate in weekly class sessions in groups of 10-20 trainees.

Instruction is presented in a co-teaching arrangement. One trainee from the cohort group is hired by Indiana University to serve as an on-site field trainer for the courses. The field trainers act as group facilitators for the sessions, coordinate class activities, oversee practice exercises, and monitor group projects. A university-based faculty member administers and supervises the course, organizes the content of the classes, prepares training materials, and participates in class presentations by means of the audiographic link.

Before each course meeting, graphic materials are prepared for the class presentation. Instructional materials can include notes, outlines, figures, charts, or other graphics. The materials are scanned, sequenced, and stored as a slide presentation that corresponds to the topics covered in the class. Materials are transmitted to the regional site through the audiographic network and previewed by both instructors. Plans are made for incorporating materials into instructional lessons, group discussions, exercises, and other class activities, and preparations for the session are finalized.

An audiographic link-up between Indiana University and the field sites is established and maintained during the class meetings. The link-up permits two-way voice and graphic interaction during lecture and discussion activities. The software communication program allows both instructors to control the slide presentation and display course materials to conform to the pace of instruction. Depending upon the structure of the class, responsibility for presenting materials and monitoring discussions can be assumed by the university instructor, the field instructor, or shared by both.

Additional in-class exercises, small-group and large-group projects, and similar activities have been devised for each class session to promote application and synthesis of concepts. The audiographic network between sites permits trainees to scan and transmit samples of their work to the university instructor for review, feedback, further discussion, and evaluation.

At the completion of the class meetings, the university and field instructor use the audiographic network to review the day's activities and assess the progress of the trainees. Preliminary plans for the next course meeting are then made.

Use of Audiographics in Supervision of Practicum Experiences

Once trainees have mastered the basic principles that form the content of an academic course, they participate in a semester-long practicum experience. The function of each practicum is to give trainees opportunities to apply the principles to the demands of their teaching positions and to learn to adapt them to the conditions of their work environment. Thus, trainees complete practicum projects within the context of their own classrooms and school buildings. Projects are collaborative in nature, and trainees work together on them as school-based teams.

Practicum activities are coordinated and supervised by the university-based instructor. To assist in carrying out and monitoring practicum experiences, the university instructor prepares written guidelines for the projects. The guidelines include a sequential listing of component steps involved in completing the projects, description of outputs that trainees produce for the activities, and descriptions of supervisory feedback that trainees can expect to receive for the steps. The instructor also assists planning and scheduling practicum activities and oversees and evaluates each group's performance. Since activities completed during practicum correspond to procedures addressed in the coursework, trainees and their supervisor know beforehand what their projects will accomplish.

Audiographic technology is used during practicum experiences to enhance monitoring and supervision of practicum activities in the off-campus school sites. By means of the audiographic network, trainees can receive feedback about their performance with very little delay. For example, about every week, the university-based instructor establishes a voice and graphic link-up with the school-based teams and reviews each group's accomplishments. As progress reports are presented, documentation of work completed can be scanned and transmitted to the instructor. Transmissions can take the form of hand-written notes, student work samples, charts, tables, or other graphic formats that trainees use to report their progress. Materials can be displayed on the monitor or printed if hard copies are needed. Thus, while trainees make verbal reports of their activities, the instructor can review the documentation and indicate areas that require elaboration or clarification. Approval, suggestions for revisions, editorial comments, or other feedback can then be provided by the supervisor. Audiographics allows both trainees and the instructor to highlight features in the materials using the computer keyboard, a graphics tablet, or a mouse. Notes can be added to the materials or they can be modified, re-scanned, and re-transmitted. Thus, documents can be revised on-line, if desired.

Important Features of Audiographic Technology in Distance Education of Teachers

Audiographics can give teachers in rural communities access to university level training when more conventional forms of distance education are not available to them. For example, many school corporations in southern Indiana cannot easily be reached by video broadcasts from Indiana's educational television system. Through audiographic technology, Indiana University can network with schools in these districts using local telephone service. There are several other features of audiographic technology that make it appealing for distance education of teachers. Some of the features described below are adapted from a report on the use of audiographics in the corporate environment prepared by Chute and Balthazar (1987).

1. Audiographics is a user-friendly technology. Establishing voice and graphic connections between sites can be easily and quickly learned. By following a few simple commands, the major functions of scanning, storing, and transmitting images are quickly performed. The ease of using audiographics is particularly important in teacher training activities because many school personnel have neither the time nor the expertise needed to become proficient in using more complex communication systems.

2. The equipment that supports audiographics is portable and easily moved to different sites. Given the current costs of audiographic technology, it is likely that teacher training institutions, rather than local school districts, will purchase the equipment. This will be true especially for those universities serving small rural school corporations. Portability of the equipment and its availability for use in multiple training sites can improve the cost-effectiveness of audiographics.

3. Scanned images can be stored on both hard and floppy disks. Thus, a large number of graphic images can be prepared, scanned, sequenced, and stored for use at a later time. Slide presentations for several classes or an entire course can be set-up well in advance. By storing class materials on floppy disks and mailing disks to the field sites, on-line transmission costs can be significantly reduced.

4. Instructors at both the origination and receive sites can control all equipment functions. For example, university-based faculty can scan, sequence, and load a slide presentation for a class onto the computer at the field training site using local telephone lines. University or field instructors can proceed through the slide presentations at a pace that suits the progress of the class.

5. Audiographics enables users at each site to interact with one another on both an audio and visual dimension. Not only can university-based instructors transmit images to trainees, but trainees can transmit work samples, outlines, reports, and other materials to instructors for review and feedback.

The ability to scan and annotate images on-line at each site makes audiographics particularly useful for monitoring class discussions and small group activities, supervising individual practicum projects, and other training experiences where two-way communication of graphic materials is an essential element of instruction.

6. During class meetings audiographics provides users with several options for communicating with one another. Users at any site can give responses verbally or annotate images by means of the computer keyboard, graphics tablet, or mouse. This feature allows instructors to highlight key points in a presentation or make corrections and modifications in the course materials. Trainees can make responses to presentations verbally or by using the computer keyboard. For example, during a class lecture by their instructor over a speaker phone, trainees can cue questions or indicate the need for clarification of key points without interrupting the verbal presentation. Furthermore, either the instructor or trainees at another site can use the keyboard or graphics tablet to provide answers to questions or give examples of concepts.

7. Class presentations can be given simultaneously at multiple sites. Although the courses offered by Indiana University currently use a point-to-point network with field sites, it is possible to offer them in a multi-point configuration. This feature of audiographics is especially beneficial in situations where it is not practical for trainees from several schools to commute to a central training site. By using a conferencing bridge, such as the AT&T Alliance Network, up to 59 separate sites could be simultaneously linked together for class presentations.

8. Audiographics also can be used for individually paced instructional activities or for later review of course materials. For example, in situations where trainees are unable to attend class presentations, lectures can be audiotape-recorded and graphic materials stored on disk. Trainees can listen to the tapes and review course materials at their own convenience.

Concluding Statement

Audiographics is an interactive technology that offers the capability of simultaneously sharing voice and graphic images. Within the context of distance education of practicing teachers, audiographics gives universities considerable flexibility in organizing and offering a wide range of training experiences adapted to the needs of rural schools. Audiographics can assist universities and school districts in creating partnerships, in carrying out staff development activities, and in devising new models of training. □

References

Anderson, C. Televised Instruction Delivery Systems: Talkback in Oklahoma for Special Needs Education. *Rural Special Education Quarterly*, 1989, 9, 29-32.

Chute, A., and Balthazar, L. A Review of AT&T National Teletraining Center Research and Development Projects. Cincinnati: AT&T National Teletraining Center, 1987.

Condon, M., Zimmerman, S., and Beane, A. Personnel Preparation in Special Education: A Synthesis of Distance Education and On-Campus Instruction. *Rural Special Education Quarterly*, 1989, 9, 16-20.

Egan, M., McLeary, I., Sebastian, J., and Lacy, H. Rural Preservice Teacher Preparation Using Two-way Interactive Television. *Rural Special Education Quarterly*, 1988, 9, 27-32.

Helge, D. The State of the Art of Rural Education. *Exceptional Children*, 1984, 50, 294-305.

Helge, D. Images, Issues, and Trends in Rural Education. National Rural Research and Personnel Preparation Project, Murray State University, Murray, KY, 1983.

Helge, D. Final Project Report of the National Rural Research Project, National Rural Research and Personnel Preparation Project, Murray State University, Murray, KY, 1979.

Keegan, D. On Defining Distance Education. In D. Stewart, D. Keegan, and B. Holmberg (Eds.), *Distance Education: International Perspectives.* New York: St. Martin's Press, 1983.

Knapczyk, D. Design and Supervision of Field-based Practicum Experiences in Rural Communities. *Proceedings of the American Council on Rural Special Education.* Bellingham, WA: American Council on Rural Special Education, 1989.

Lehr, J., and Harris, H. *At-Risk, Low Achieving Students in the Classroom.* Washington, DC: National Education Association, 1988.

McConnell, D. Sharing the Screen. *Media in Education and Development*, 1983, 59-62.

O'Connor, N., and Rotatori, A. Providing for Rural Special Education Needs. In A. Rotatori, M. Banbury, and R. Fox (Eds.), *Issues in Special Education*. Mountain View, CA. Mayfield, 1987.

Pelton, M. *Staff Development in Small and Rural School Districts.* Arlington, VA: American Association of School Administrators, 1983.

Slavin, R., Karweit, N., and Madden, N. *Effective Programs for Students At-Risk.* Needham Heights: Allyn and Bacon, 1989.

Treadway, D. *Higher Education in Rural America.* New York: College Entrance Examination Board, 1984.

Williamson, R. The Cyclops Audiographic System. *Media in Education and Development*, 1983, 63-65.

The Design and Application of a Distance Education System Using Teleconferencing and Computer Graphics

Bill Winn, Barry Ellis,
Emma Plattor, Larry Sinkey, Geoffrey Potter

Another Application of Computers

To most people, the use of computers in education means either computer-assisted or computer-managed instruction. This article reports an experiment in another application of computers—the use of computer graphics for the direct instruction of students in distance education courses. Rather than stressing the interactive capability of instructional computer programs, we took advantage of a computer's ability to store and provide instantaneous access to print and graphic material, and to send that material over phone lines to students at remote sites, while leaving the interactive aspect of instruction to instructors teaching over an audio-teleconferencing network.

We believe that the system we have developed to accomplish this has considerable potential for instruction, whether in conjunction with human teachers or, at some future time, coupled with the more familiar CAI and CMI. In this article we describe the system, the instructional roles of the graphics we used, and how we developed two courses which we used to assess the system's effectiveness. Some of the results of that assessment are presented.

System Description

Teleconferencing

The teleconferencing system at the University of Calgary uses the telephone to connect a sound studio on campus, from which the instructor teaches, with up to twenty remote centers throughout the Province of Alberta (for a description of the system, see Ellis and Keenan, 1983). The heart of the system is a piece of equipment known as a "Bridge," a device controlled by a microprocessor through which a signal from the studio is amplified and sent to the centers. It also allows the centers to talk back to the instructor in the studio and to talk to each other. Centers can be connected to work independently of the instructor, allowing the electronic equivalent of "small group activities" in class.

In each center is a "Convener." This device contains a speaker and some switching circuitry. It is connected to the telephone system either directly or by means of an acoustic coupler. Attached to the convener are a convenient number of microphones, which students can speak into at any time simply by pressing a "talk bar." The student is heard by the instructor and in all the other centers. This makes the system fully interactive. As well, the system can be switched to "broadcast" mode, which turns the system over to the instructor exclusively.

Typically, an instructional session will consist of presentations by the instructor, during which students may interrupt and ask questions; discussions and activities within and among centers; and reports about homework assignments completed between classes. The system is not used predominantly for instructor presentations.

NAPLPS/Telidon

The NAPLPS (North American Presentation Level Protocol Syntax) videotex system, like its predecessor, Telidon, around which our system was built, comprises a communications protocol and a decoder for displaying text and graphics in eight colors, eight textures, and eight tones of grey (see Ruggles, Anderson, Blackmore, Lafleur, Rothe and Taerum, 1982, for a fuller description of the system, and O'Brien, Brown, Smirle, Lum, and Kukula, 1982, for a technical description). It is presently gaining acceptance for use in business, industry, government, and the home. Anyone with a decoder and some form of CRT display screen, such as a television set, may access "pages" of text and graphics from a database by means of a keypad or keyboard. Pages can be added to the database at a workstation consisting of a keyboard, digitizer pad, and two CRT displays, one showing the page as it is created, the other displaying menus of commands for page creation, file handling and editing.

Our Telidon software and database were supported by a VAX 1170 computer running UNIX. The software was written in C. It consists of pro-

Bill Winn is at the College of Education, University of Washington, Seattle, Washington. Barry Ellis is at Olds College, Olds, Alberta, Canada. Emma Plattor and Larry Sinkey are at the University of Calgary, Calgary, Alberta, Canada. Geoffrey Potter is at the University of Victoria, Victoria, British Columbia, Canada.

grams that are easy for the novice to use for page creation, editing, display, and file management.

A Telidon page can contain both text and graphics. The text can be of three sizes plus double height and width, and in any of Telidon's colors. (More recent versions of NAPLPS provide the user with a far greater choice of colors and other graphic components.) The graphics consist of "objects," which in turn are made up of "elements." The elements are defined by the system. They consist of different types and thicknesses of lines, geometric shapes (circles, arcs, rectangles, etc.), and polygons which allow the user to create shapes as complex and irregular as are needed. Thus, pictures with a fairly high degree of complexity can be created. Any shape or polygon can be left open or can be filled with any color or level of grey. Elements may be duplicated, rotated, reflected, or moved once they have been placed on the page.

Objects, the next level of complexity, are made up of elements. The object "house," for example, might be made up of the elements rectangle, triangle for the roof, another rectangle for the chimney, with an irregular polygon depicting the trees in the yard. Like elements, objects can be duplicated, moved, rotated, and reflected. They may also be stored so that they can be used again on another page.

Finished pages are stored in a tree structure as files. They are recalled for display by keying in a number that points to the branch of the tree and the place of the page in the branch. Pages can be edited, and moved to other places in the database.

Interfacing

Interfacing Telidon to the teleconference system required attaching a Telidon decoder and monitor to the convener in each center and getting the Bridge to accept and distribute signals from the VAX computer. The latter was achieved through the use of appropriate modems, although it is still not possible to send voice and data at the same time. (The incorporation of a multiplexing device that would allow this was too costly for this initial project.) Thus, the instructor has to stop talking while a graphic is being sent to the centers. A manual, and recently a tone-activated switch, to turn the system over from voice to graphics and back again, have been developed. The manual switch is cumbersome to use, requiring that a switch be thrown in the control room and in each center whenever a page is to be sent. The tone-activated switch has overcome this difficulty, though it was not in place for this particular project. Both switches, incidentally, enable the transmission of any data, including ASCII code and equipment control signals, and Telidon text and graphics.

Instructional Application

Background

The Telidon-Teleconference system has been extensively tested and evaluated. Over a period of three years, from 1982 to 1985, it has undergone considerable modification, both in hardware and computer software. Its instructional effectiveness has been assessed in courses within the Faculty of Education's Inservice Program for teachers. A course in teaching teachers how to teach grammar has been offered twice using the system, and a course on classroom management for language arts instruction has been taught once. Both courses were "quarter courses," involving twenty hours of instruction per week for seven weeks during the evening.

Role of Computer Graphics

Both of these courses lent themselves well to graphic support. Research on the instructional use of graphics (Dwyer, 1972, 1978; Fleming and Levie, 1978; Winn and Holliday, 1982), and on their potential value in our system (Plattor and Winn, 1983), has identified a number of advantages to using pictures, charts, and diagrams in instruction. (In particular reference to Telidon graphics, Mills, n.d., has provided a thorough analysis concerning their design and effectiveness in communication:)

The first advantage of using graphics in instruction lies in their ability to illustrate. When new concepts are introduced to students, it helps if the critical attributes of those concepts are presented as realistically as possible (Merrill and Tennyson, 1977). In both courses, we used realistic presentation of attributes to teach concepts. For example, sentence-combining techniques were illustrated by graphics in which depictions of what was occurring in each sentence were successively added so that a composite illustration, corresponding to the composite sentence, was created.

A second advantage of graphics is their ability to highlight important material (Winn, 1981). Graphics can draw attention to some feature of the content by making it stand out against the other information on the screen. Thus, contrasting colors, type sizes, and unusual layout, combined with such techniques as underlining, circling, enclosing text or drawings in boxes, arrows and blinking, were used for cueing. (All of these techniques are easy to use with Telidon.) For example, in compiling a list of classroom management strategies, the newest item would be presented in a bright color, but would change to a muted color when the next item was added. Boxes would be placed around parts of speech in a sentence. Important words would blink for a few seconds.

Another useful way of considering the instructional functions of graphics is to look at how they illustrate temporal and spatial structure. To begin with, a graphic can be built up a piece at a time to show a sequence of events. Thus, the transformation of two simple sentences into a more complex one can be illustrated a step at a time. Telidon has the facility to display "overlays" just as an overhead transparency can. This is done by displaying a new page without removing the one that is already on the screen. In this way, communications models for classroom management were built up a section at a time, so that not too much new information was presented at once, while the way the whole process fit together emerged.

Graphics, particularly diagrams, are good at showing spatial layout, and thus the structures and arrangements of things (Bartram, 1980; Duchastel and Waller, 1979). This is because they can illustrate both actual and metaphorical distances. In the former case, diagrams of classroom layouts can illustrate how management may be facilitated through varying the literal proximity on the screen of teacher to student, student to student, and student to equipment, shelves, room dividers, and so on. In the second case, a diagram might show a word, naming a concept, in close proximity on the screen to another word, naming another concept. The effect of this for the student is that the two concepts are thought of as being similar in some way because of their close physical proximity (Fleming and Levie, 1978, p. 70). Charts of different types of sentence or of different types of writing that children might do, where the proximity of words to each other varies, are examples of how we used this instructionally.

Course Development

The two courses were prepared using acknowledged procedures for instructional development (see Dick and Carey, 1985). Two aspects of the project were somewhat exceptional, however, and deserve mention. First, because the delivery system was being created at the same time as the courses, we had the opportunity to involve specialists in computer software and hardware in the course preparation process. This meant that the system we have created is to a large degree specifically designed for instruction. Second, we gave particular attention to the identification, design, and formative testing of graphics to accompany the spoken part of the instruction. The subject matter specialists suggested where they thought a graphic might be useful. A sketch was prepared, often on the spot, to which the content specialists reacted. A

revised drawing was given to the page creator, and the resulting Telidon page was critiqued by the subject specialists. Revisions were made, or the page was discarded. Then all of the pages in the database were critiqued by groups of students chosen from the target population. This gave us information which led to further revisions or deletions. What emerged was a database of carefully designed and tested Telidon pages, course manuals in which every presentation, class activity, and assignment was described in detail, and student handouts. It should be noted that both courses were co-taught, and that the teaching of each session was shared by at least two instructors.

System Effectiveness

A major goal of the grammar course was that the students should apply what they had been taught to their own teaching. Indeed, some of the assignments required them to try techniques they had been taught during the week between two class sessions and to report back to the course participants on their success. Because of this, the main thrust of the evaluation of the grammar course was on the degree to which the students applied what they had learned. Our concern in evaluating this course was simply to find out if the system worked. No systematic attempt was made to assess the impact of the computer graphics separate from the system as a whole.

At the beginning of the grammar course, the students were given a pretest that assessed their understanding of twenty-two major concepts and skills presented in the course and the degree to which they already applied them. During the course, each student kept a diary in which was noted the frequency of the student's use of the same concepts and skills. The analysis of the "use profile" that these diaries provided allowed us to determine whether or not the concepts taught in the course were being applied. In addition, students completed questionnaires, distributed three times during the first three months of the next academic year after the course ended. The questionnaires assessed the degree to which students continued to apply the twenty-two concepts and skills.

Based on data from the fifty-five students enrolled in the course, a clear increase in the application of the concepts and skills was evident. On the pretest, thirty-six students indicated that they recognized four or fewer. Only three students recognized all of the items. As far as use was concerned, fifty students made regular use of six or fewer of the items. This indicates that, by and large, the students had a low entering knowledge of the course content and applied it even less.

Examination of the diaries revealed that, with

the exception of five students who were not teaching class while the course was being offered, the majority applied the course content consistently. The use profiles varied greatly among the students, as one would expect. However, in the case of forty-four of the fifty students with access to classrooms, an increase in the use of the concepts and techniques taught in the course was observed.

Results from the three questionnaires distributed from September to November showed that students continued to use what they had learned well into the next school year. For September, on average, each student made use of forty-four percent of the content. In October, this increased to sixty percent, and in November it was forty-eight percent. Broken down by frequency of use, it appeared that more content was used "occasionally" than "regularly," while little was used "frequently." However, given the fact that the content of the courses that the students had to teach did not overlap completely with what they had been taught in our course, these results suggest a satisfactorily high rate of use.

In the case of the classroom management course, it was less likely that the techniques and concepts could be applied as immediately. Many of the techniques taught in the course required fairly radical changes in classroom arrangements and procedures, which might be impossible to implement without a change in attitude and sanction by school authorities. It was therefore decided to conduct a more traditional evaluation. Before each session, the thirty-six students filled out a brief questionnaire on which they rated on a five-point scale their degree of familiarity with and the degree to which they used the major concepts and techniques for that particular session. After the session, they filled out another questionnaire, assessing how well they understood the concepts, and how much they thought they would be able to apply the techniques in their classes. The post-session questionnaire also asked questions about the role of computer graphics and the success of the Telidon-teleconference system in that particular session.

When pre- and postsession ratings of students' knowledge of the concepts and techniques taught in each session were compared, statistically significant increases were found ninety percent of the time. In similar comparisons of when students thought they would be able to apply what they had learned, significant increases were found in seventy-four percent of the cases.

The average ratings for the statements concerning the effectiveness of Telidon suggested that after each session the students were very positive towards the computer graphics and to the way they had been used in concert with the other elements of the system. It appeared that our attempts to use the graphics for illustration, showing sequences, structures and spatial relationships were successful.

Answers to this questionnaire also indicated that students took few notes from the computer graphics at the beginning of the course, but took progressively more as the course went on. This suggests that they came to trust the information presented in the graphics as they got used to them. However, other answers told us that, although our system was better than most others they had encountered when taking off-campus courses, it was still not preferred over learning in a traditional classroom from an instructor whom they could see.

Finally, the students really liked the pages that had pictures in them. We frequently used cartoon characters as part of descriptive and illustrative graphics. Apparently, these heightened interest and made it easier for students to understand the ideas and techniques that the graphics described.

Conclusion

Frequently, technological innovations are placed in educational settings and used without the planning, design, assessment, and revision that are necessary if they are to work well. In our project, we emphasized these activities, and it paid off. Our rather thorough evaluation showed that the system succeeded in teaching the students the ideas and techniques we wanted them to know. It did not do this any better, nor any less well, than other systems. But that is not the point. It worked.

Looking more carefully at the particular contribution made by Telidon, we found that the graphics were generally successful in the roles we intended for them—illustration, drawing attention to important material, and explaining spatial and temporal relationships, both literal and metaphorical. Students liked the graphics, thought they were useful, and trusted the information they gave. Information about problems with particular pages was useful to us for making even further revisions to the graphics, but has not been discussed here. Other, more general, guidelines for the design of computer graphics have been presented elsewhere (Winn, Plattor, and Loosmore, 1985).

One final observation. Our system could have been a lot fancier and a lot more expensive. We could have multiplexed the signal or used two phone lines into each center so that we could talk and send graphics simultaneously. We could have placed intelligent terminals in each center so that pages could have been downloaded ahead of class and displayed locally. We could have used "real-time" graphics creation, or two-way video. We chose not to, believing that it was better to develop a fairly simple and inexpensive system that we could design and test thoroughly rather than to attempt something that might ensnare us in technical difficulties. (Even with our simple system, we had enough of these!) As a result, we believe that our system, or a modified version of it, would be easy to set up in any setting with access to fairly standard facilities like telephones and a computer.

The success of this, or any subsequent project, relies not on sophisticated hardware, but on the skills of the designers, programmers, subject specialists, and instructors. Our experience has shown that these skills are not difficult to acquire, provided that the project is not over-ambitious. □

This project was completed while the first and second authors were at the University of Calgary. It was supported by a grant from the Innovative Projects Fund, administered by the Government of Alberta's Department of Advanced Education. We are grateful for the support and encouragement given to our project by this agency and by many of its employees.

References

Bartram, D.J. Comprehending Spatial Information: The Relative Efficiency of Different Methods of Presenting Information About Bus Routes. *Journal of Applied Psychology*, 1980, *65*, 103-110.

Dick W., and Carey, L. *The Systematic Design of Instruction* (Second Edition). Glenview, IL: Scott, Foresman, 1985.

Duchastel, P.C., and Waller, R. Pictorial Illustration in Instructional Texts. *Educational Technology*, 1979, *19*(11), 20-25.

Dwyer, F.M. *A Guide for Improving Visualized Instruction.* State College, PA: Learning Services, 1972.

Dwyer, F.M. *Strategies for Improving Visual Learning.* State College, PA: Learning Services, 1978.

Ellis, G.B., and Keenan, T. Microcomputers, Videotex and Educational Teleconferencing. In C. Keren and L. Perlmutter (Eds.), *The Application of Mini- and Microcomputers in Information, Documentation and Libraries.* New York: Elsevier, 1983, 295-308.

Fleming, M.L., and Levie, W.H. *Instructional Message Design: Principles from the Behavioral Sciences.* Englewood Cliffs: Educational Technology Publications, 1978.

Merrill, M.D., and Tennyson, R.D. *Teaching Concepts: An Instructional Design Guide.* Englewood Cliffs: Educational Technology Publications, 1977.

Mills, M.I. *A Study of the Human Response to Pictorial Representations on Telidon.* Ottawa: Government of Canada, Department of Communications (n.d.).

O'Brien, C.D., Brown, H.G., Smirle, J.C., Lum, Y.F., and Kukula, J.Z. *Telidon. Videotex Presentation Level Protocol: Augmented Picture Description Instructions.* Ottawa: Government of Canada Department of Communications, 1982.

Plattor, E., and Winn, W.D. An Integrated Audio and Graphic Distance Education System, and Its Potential for Teaching Chemistry. *Chemistry in Canada*, 1983 (October), 12-15.

Ruggles, R.H., Anderson, J., Blackmore, D.E., Lafleur, C., Rothe, J.P. and Taerum, T. *Learning at a Distance and the New Technology.* Vancouver: Educational Research Institute of British Columbia, 1982.

Winn, W.D. The Effect of Attribute Highlighting and Spatial Organization on Identification and Classification. *Journal of Research in Science Teaching*, 1981, *18*, 23-32.

Winn, W.D., and Holliday, W.G. Design Principles for Diagrams and Charts. In D. Jonassen (Ed.), *The Technology of Text* (Volume One). Englewood Cliffs: Educational Technology Publications, 1982, 277-299.

Winn, W.D., Plattor, E., and Loosmore, J. Designing Computer Graphics: An Experiment in the Graphic Enhancement of Audio Teleconferencing. Paper presented at the annual meeting of the American Educational Research Association, Chicago, April 1985.

Teletext: A Distance Education Medium

Aliza Duby

Introduction

"It is an interesting time for educational broadcasters, because a range of newly articulate audiences now want from them more responsiveness to *specific needs*, more accessibility and involvement, more opportunities to express and to hear expressed a variety of ideas, more representation for minorities and more *choice*." (Innes, 1985, p. 143)

Educational Teletext is the modern age's answer to those new audiences, because Teletext is a mass medium that caters to specific groups and needs.

Teletext gives the learner the opportunity to interact, to practice skills, to pace learning according to ability and inclination; it gives him access to data and feedback at will, and also permits him to build his own learning package and to have control over his learning. In these respects Teletext's advantages serve not only the learners but also the professionals and other specialized people (Innes, 1985; Ettema, 1984).

The information revolution and information technology are stepping up the demand for education, and increasing the potential economic and social penalties for not responding to those demands (U.S. Congress, Office of Technology Assessment, 1982).

So, in order to cope in the 21st century, educational institutions such as universities and libraries should harness the innovations of our times to their educational systems.

Teletext Educational Technology

Teletext is a hybrid-resulting from the mating of computer controlled programs and the television screen. Teletext information is delivered on unused television signals, so it can be broadcast simultaneously with TV, but it is used mainly as a separate, unique medium. The Teletext frames (called Teletext pages or Teletext screens) are transmitted perpetually and reiteratively in sequential order.

Aliza Duby is a Project Leader with the South African Broadcasting Corporation, Johannesburg, South Africa.

At the touch of a button, the TV screen becomes a Teletext screen. The user can call any page that he wants onto his screen at will. Usually he has to wait a few seconds for the page that he has called to be displayed on his receiver. Once "captured," the page is stored in the receiver's local electronic memory, and this page can be displayed for an unlimited period, to be picked up at any pace. It can be run forwards, backwards, be stopped or frozen. The text can be enhanced with graphic illustrations (Zoroczy, 1981).

The teletext pages are stored remotely, where they can be kept *continuously up to date*. Each page is identified by a numerical reference. The teletext information is classified and indexed and can be selected and displayed at will. The entire data base is continually re-transmitted.

As to printing, the system reduces the normal delays in centralized printing and distribution (Woolf, 1981).

Teletext is a pseudo-interactive medium; some of its qualities give the user more involvement in the learning process than do other educational media, e.g., the "Reveal and Conceal button." With this technique, the answer to a question does not appear on the screen until the response key on the remote control is pressed. The technique can be used for programmed learning, study quizzes, self-achievement testing, and obtaining additional information. This response technique is important for individual learning, in which each student can control his own learning (Issing, 1986).

Consequently its main *technical advantages* are the following:

—Different programs can be watched at the same time by different individuals.
—It possesses the logical control and flexibility of computer software.
—Users determine their own routes: programs are selected by users.
—The system has color adaptability; it employs the three fundamental colors and combinations of them. Either letters, graphics, or the background can be in different colors.
—Proceeds at one's own pace.
—Can be used at will.
—For unlimited duration.
—In one's own environment.
—Can be watched by an individual and/or a group.
—No special qualifications are necessary for the use of the equipment.
—A modern, cost-effective data-transmission network.
—Constant and fast updating of the software.
—The ability to deliver information instantaneously.

—Highly cost-effective transmission medium, relying on modified but existing equipment. (The fact that the teletext signal travels on free TV signals makes the cost of transmission per user negligible.)

—Direct compatibility with computing facilities (via a mass-producible adaptor). Even though teletext is not a fully interactive medium, the software that it carries may be down-loaded and used interactively.

—Teletext in its present form is restricted to a static visual presentation, but can, in principle, be extended and accompanied by sound. (This would presuppose the availability of broadcasting or cable channels.)

—Captions added to TV programs could enhance the educational impact of TV.

—The ability of the user to select information pages and to reveal information on demand makes it an ideal self-learning medium.

—However, more widespread availability of two-way cable would turn Teletext into a truly interactive medium.

Hosie (1987) suggests that "Encouraging individual routes through information will assist students to become more actively involved in the learning process and thus have better control over learning strategy."

Teletext gives the self-motivated user the opportunity to have control of his own learning. It gives the user the chance to practice skills; to pace learning according to his inclination; to have feedback; to improve his own learning management; and, most importantly, it gives him access to information at will and according to his choice. Consequently, Teletext is perhaps the ideal medium for continuing education and for individual learning.

With regard to the recall of information, it was found (Edwardson et al., 1985) that because reading is a more active process than listening, seeing the written word is more memory effective than hearing the spoken word. Also, a message that is accompanied by graphics is retained for longer than one that is not. It was also found that a redundant message conveyed by Teletext captions riding on TV is remembered better than an unredundant message (when the crawls convey another message).

We can conclude that it is advisable to deliver a message which conveys information to be remembered by Teletext rather than, for example, by a TV talking head or by radio. However, it should be remembered that Teletext is a medium for literate people who are used to reading. Teletext is therefore suitable for the continuing education of the distance learner.

Although Teletext text is constantly updated, it is a "non-real-time" medium, as are newspapers and computers (but not TV or radio). From this point of view it has advantages over these media, because as mentioned above one can use Teletext at one's own time and pace, and one gets newly updated information. One can obtain more information, and have control of the information processing. One selects the content one wants to attend to, the quantity, and its order of appearance. One can also use the concealed information to get some feedback or more information. The technology can compensate for some of the major weaknesses attributed to instructional television regarding learner control, such as lack of opportunities for reflection, selection, and repetition (Bates, 1981).

Teletext is a computerized device, and as such employs Computer-Assisted Learning strategies, which are to the advantage of the distance learner.

—Changing and updating the Teletext script is easy; as easy as typing a letter. That makes Teletext a most effective medium for the learning of volatile subjects such as the economy, political science, etc.

—The same quality of the medium makes it an ideal modern educational tool, as formative evaluation results can be easily interwoven in program development or adaptation.

—It is a limited medium from the visual and audio points of view, so its content has to be selected carefully.

—Concise and effective courseware makes the work appealing to the student, and motivates him to use the courseware without frequent assistance from the instructor (Suzuki, 1987). The suggested approach will facilitate employing effective motivational strategies such as menu-driven structure, instructional feedback, and the congruency among the components of the courseware (Gallegos, 1987).

—Instructors provide effective instructions for each learner, e.g., this is done by using the command "go to page" after each response, and/or by using the response technique.

—It can also provide pages geared to different knowledge or competency levels, thus producing adjustment to individual needs (Tidhar, et al., 1985).

It has been recognized that the cost of Teletext pages and the limit on the amount of text that can be displayed on a single frame suggests that Teletext is not particularly useful for detailed descriptions. Instead it is good for providing an overview of the areas of a subject that could prove useful to an instructor preparing a lesson and wanting to keep up with current developments. Teletext can also provide access to information that would otherwise not be available and thus broaden the data base (Thompson, 1982).

However, Teletext's main function is as an independent medium because it presents a wide range of learning materials in effective, attractive, and entertaining ways and is designed to meet the individual needs of the distance learner.

Teletext: The Medium for Distance Education

Dhanarajan (1986) summarized some relevant characteristics of distance education:

—Separates teachers and learner.
—Conducted by educational organizations, which distinguishes it from private study.
—Make use of technical media.
—Teaches people mainly as individuals.

It follows that Teletext is a distance education medium *par excellence*.

In remote areas and in areas with burgeoning populations, Teletext could become the means of bringing education to the people. Teletext would permit learning at home when there is no on-site educational institution or where there is a dearth of qualified teachers. Some countries have experimented with this kind of special distance education service. For example, "Prestel Farmlink" (Britain) is intended for the farmer (in operation since April, 1984). It is a comprehensive farm management service and is being progressively extended nationwide. This program is "designed to help farmers make the most of their experience and judgment, cutting down on time spent looking up for information" (Shearer, 1984).

Teletext could also accommodate the needs of *mentally and physically handicapped* children who attend special schools, as well as those handicapped people who are not able to attend schools. For example, it is the only mass medium available for the deaf. Actually, Teletext was initially developed for the deaf in Britain in 1979 (Brown, 1974).

The system can also accommodate the needs of adults who are interested in *continuing* their education. They could watch programs on subjects such as preparation for matriculation examinations, general knowledge, professional guidance, or computer programming. For example, "Prestel Micronet 800" offers a range of services for *computer users*; Teletext pages store computer programs, which can then be recorded off-air directly onto a microcomputer.

It could offer a *Guidance program* that would help people to cope with the changing world; to meet the constantly changing conditions of the working world; to cope with adolescent sons and daughters, etc. Career information, too, is seen as an area for distance education. Not only could it provide additional guidance in the selection of a career, but also specific data about job opportunities.

Teletext can change and enrich functions of the *library*. A special library data base could be centralized by Teletext, e.g., national library catalogs could be transmitted on Teletext. That would save librarians and readers time and money. At the moment most catalogs are on microfiche, and only a special librarian in a library environment has access to the catalog and can request the information needed by using the postal inter-library loan system. With Teletext this lengthy interaction process for obtaining information could be dramatically simplified.

Librarians involved in a field research program on the use of Teletext in the library (Badzik, 1983) list the system's advantages: it provides library users with up-to-date information; draws more people to the library; fills gaps in existing information coverage; provides a quick way of retrieving data; broadens means by which information is available, presenting it in a novel manner; and has the potential to relieve the staff's workload.

Other benefits of Teletext in the library include: reducing the cost of in-house and inter-branch communications because it eliminates paper; gaining a new medium of publicizing services, new acquisitions, and programs that would cost less than direct-mail literature; providing an aid to answering reference questions; and expediting orders for services such as books by mail.

For industry the advantages of the medium for training are obvious. It could be used in the company training center at will, for as long as would be needed by the individual employee as well as by a working group. It could be used as a complete program, or as a supporting program to be controlled by the company supervisor. For example, in England, Barclays Bank has been using videotext for *staff training* for several years. They have 60 courses consisting of an average of 450 frames, with a running time of 30 minutes to 3½ hours (Brown, 1986).

The use of Teletext is seen as a way of introducing information technology to the distance learner and of enabling him to become familiar with information concepts and information skills such as information, classification, retrieval skills, the planning of logical searches, and the use of indexes (Thompson, 1982).

Further Research

If we are to embark on this venture—to use Teletext for distance education—further research to explore the range of its applications should be undertaken. The following are some suggestions in this direction.

- —Retrieval of information: The most useful and thus the most used facility for distance education is the library. Library functions should therefore be researched in order to find out how Teletext can be used to improve library services for the distance learner.
- —There are wide-open areas such as Viewdata, Interactive Teletext, specialized Teletext for "open universities," services for the deaf, in-service studies, etc.
- —Teletext is already used in at least 30 countries for innovation and education. Great Britain and Canada are the leading countries; the USA and Japan follow them, and a further 26 countries have begun to experiment with Teletext: These are most of the European countries, Australia, Hong Kong, Maylasia, New Zealand and Singapore (Bacsich, 1986). There were trials, begun in 1985 and 1986, on international links between France and Germany, and between France and the United Kingdom. This means that in the future Teletext will probably provide us with an international data base. The possibility of interacting with an international network and so having a real distance education service should be investigated.

Conclusion

Teletext is a new medium, which has special characteristics. As such it is an addition to the existing educational media, so it does not threaten the status of either the teacher, Educational TV, educational radio, or even the textbook. We should take advantage of the unique qualities of Teletext in order to solve unsolved educational problems, such as educating at a distance. □

References

Appleyard, R.K. Opening of the Conference. In C. Verniby and W. Skyvington, *Proceedings of the Videotex in Europe Conference*, Luxembourg, July 1979.

Bacsich, P.D., and Castro, A.S. A Hitch-Hiker's Guide to the World of Videotex. Paper presented at the *Seminar on the Future of Videotex in Tertiary Education*, Sydney, July, 1986.

Badzik, S.K. Videotex: Blessing or Bane for the "Boob Tube"? In E. Sigal, (Ed.), *The Future of Videotex*. London: Kogan Page, 1981, 123-129.

Bates, T. Evaluation on a Tight Budget. *Educational Broadcasting International*, September 1981, 113-116.

Brown, L. The Aims, Organization, and Operation of a Broadcast Telesoftware service. Paper delivered at the *IEEE International Conference on Consumer Electronics*, 1984.

Brown, L. Tele Software: Experiences of Providing a Broadcast Service, Mimeographed, 1986.

Castro, A.S. Videotex in Tertiary Education: The Missing Links—A Keynote Address. Presented at the *Asian-Pacific Conference*, Sydney, July, 1986.

Clark, D.R. The Role of the Videodisc in Education and Training. *Media in Education and Development*, December 1984, 190-192.

Dhanargan, G. The Application of Communication Technology in Distance Learning. A paper presented at the *Canada-Maylasia Conference*, Ottawa, October 1986.

Edwardson, M., Kent, K., and McConnell, M. Television News Information Gain: Videotex Versus a Talking Head. *Journal of Broadcasting & Electronic Media*, Fall 1985, *29*(4), 367-378.

Ettema, J.S. Three Phases in Creation of Information Inequities: An Empirical Assessment of a Prototype Videotex System. *Journal of Broadcasting*, Fall 1984, *28*(4).

Gallegos, A.M. Technology in the Classroom: Another Look. *Educational Technology*, July 1987, *27*(7), 15-24.

Hosie, P. Edopting Interactive Videodisc Technology for Education. *Educational Technology*, July 1987, *27*(7), 4-10.

Innes, S.M., Barnes, N., and Brown, L. New Ideas—Where Are We Going? *Journal of Educational Television*, 1985, 143-144.

Issing, L.J. Interactive Videotec: A New Medium for Education. Paper presented at the *Joint Japanese-German Symposium on Information-Oriented Society*, Tokyo, 1985.

Livingston, K. Brokering in Distance Education, Australian Style. *Media in Education and Development*, March 1987, 10-13.

Sherrington, R. *Television and Language Skills*. Oxford: Oxford University Press, 1973.

Shearer, J. Prestel: British Telecom's Public Viewdata Service. *International Journal of Micrographics & Video Technology*, 1984, *3*(2), 111-114.

Suzuki, K. A Short-Cycle Approach to Development: Three-Stage Authoring for Practitioners. *Educational Technology*, July 1987, *27*(7), 9-24.

Tidhar, C.E., and Ostrowitz-Segal, L. Teletext in Israel: A New Instructional Tool. *Journal of Educational Television*, 1985, *11*(3).

Thompson, V. Videotex in Education. *Media in Education and Development*, September, 1982, 118-120.

U.S. Congress, Office of Technology Assessment. *Information Technology and American Education*. Washington D.C.: US Government Printing Office, 1982.

Woolfe, R. Videotex and Teletext: Similarities, Differences and Prospects. *Programmed Learning & Educational Technology*, November, 1981, *18*(4), 245-253.

Zorkoczy, P. Teletext Systems. In E. Sigal (Ed.). *The Future of Videotext*. London: Kogan Page, 1981, 165-172.

Local and Long Distance Computer Networking for Science Classrooms

Denis Newman

In the conduct of modern science, collaboration is seen pervasively both within laboratories and among scientists spread out around the world. No scientist can make a contribution to knowledge without building on the work of his or her colleagues. Journals and shared databases facilitate collaboration locally and over greater distance. Division of labor on scientific problems, such that a group of scientists converge on a joint solution, is also an important aspect of scientific work. Children who are learning science will benefit as well by being treated as young scientists and being given the opportunity to collaborate (Mitchell, 1934; Slavin, 1983). Not only might joint problem-solving facilitate scientific thinking, it will certainly model the process of doing science, making clear to the students that science can be a lively and active process.

The Earth Lab project, funded by the National Science Foundation, is demonstrating new ways of using computers for upper elementary and middle-school science instruction. Earth Lab uses a local area network (LAN) to link up the microcomputers in the computer or science lab so that students can easily communicate and share information as well as work together on simulated science problems. The project is also using a telecommunications network which gives students and teachers communication access to colleagues in other schools—both in the local calling area and around the world.

A major concern in the Earth Lab project is finding ways to integrate these two kinds of computer-mediated communication. Local area networks, which use a cable to directly connect computers, usually in the same building, and telecommunication networks, which use a combination of telephone lines and long distance data communication networks, both have their particular strengths and weaknesses. Formative research on educational telecommunications, reported here, helps to define appropriate roles for these technologies. Scientists use both long distance and local computer-mediated communication. Electronic mail is used within a lab and local campus as well as among scientists spread out around the world. In the instructional as well as in the real scientific context, the distant communications are harder to coordinate and are best organized as extensions of local communicative activities. Our experience with long distance communication (Levin and Riel, 1985; Newman and Rehfield, 1985; McGinnis, 1986) suggests that a well-designed networking system is marked by a coordination between local and long distance functions (Newman, 1985).

Earth Lab: Collaboration in the Classroom

The Earth Lab project has developed curriculum units in two geography related areas—weather and climate and plate tectonics—which provide the instructional context for a test of whether the local area network technology can help teachers to create a valuable environment for learning science. Since September of 1986 Earth Lab has been operating in the sixth grade of two pilot schools.

We have created or modified several pieces of Apple II software for use in these units. These are being designed to work with the Corvus network we are using in the project but the general design principles apply to any network.

Interface. We have created a network interface which will give the teacher and students easy access to the programs, data, and text files. The interface makes it easy for a teacher to create a "workspace" for a group of students where data from a group project is kept or to create a "library" or common area for a class or for the school as a whole.

Filer. A database management system, *Bank Street Filer*, has been modified so that students can easily locate and contribute to a common database for group projects.

Writer. We have also created a special version of *Bank Street Writer*, a word processing program that is popular in elementary schools. The new *Writer* allows students to send electronic mail to anybody else in the school as well as to students at other schools. Teachers are also able to send messages to individuals or groups of children.

Simulations. A navigation simulation game, *Rescue Mission*, has been modified so that four groups of students can pilot their ships simultaneously.

We have designed science activities in which the task is divided among a group of children who use the local network technology to gather

Denis Newman is with the Center for Children and Technology, Bank Street College of Education, New York, New York.

and share their information. These activities involve a fairly high degree of coordination. For example, in our unit on plate tectonics, students will be divided into groups each of which will study a portion of a larger database of facts about where dinosaur fossils were found, the age of the fossil, and the climate in which the animal probably lived. Each group will be responsible for fossils found on a particular continent. The students, who have already studied the Earth's major climate zones in a previous unit, will then formulate hypotheses about where the continent might have been positioned so that, for example, fossils of tropical animals would have been living in the tropics. New information will then be added to the database and the different groups will compare their hypotheses and attempt to derive a general pattern of plate movements.

In another example of coordinated classroom work using the LAN, students play a navigation simulation in which four teams (each at a different computer) attempt to find each other on a simulated map space using various instruments such as a radio direction finder, radar and binoculars. The game includes a "CB" function with which the teams can send each other short text messages about their current position and direction.

Written communication is an important aspect of Earth Lab. Text is composed jointly and shared over the LAN, for example, as contributions to a class research project on weather disasters. Electronic mail is also used in the classroom. Students are encouraged, and sometimes assigned, to send the teacher a short message describing their hypotheses about the outcome of an investigation. These messages are printed out and displayed as a basis for class discussion. Since the Earth Lab activities also involve a high degree of coordination among the science teacher, the computer coordinator and the classroom teachers, the staff use the network to help get their own work done.

Formative Research: Long Distance Networks

The design of Earth Lab draws on two years of experience with communication networks among teachers and students. These earlier projects were the formative research which led to the following two design principles:

1. Simplify the access to local and long distance communications by basing it on a single writing system which is also used for conventional writing.

2. Organize activities on the long distance network which do not depend on the precise temporal coordination. Leave such activities for the local network.

The first of our formative projects is the Bank Street Exchange, a bulletin board system running on an IBM AT at Bank Street. One activity involved the exchange of pen-pal letters and short essays between elementary classrooms in New York and San Diego. In this case, there was a combination of local and long distance communication. To avoid the more expensive daytime rates on The Source, the commercial network system we were using, students made use of local communication facilities. In San Diego, the schools used the U.C. San Diego electronic mail system. In New York, they used the Exchange. Messages were portaged via The Source between the two cities. For example, in New York, children in classrooms, and sometimes from home, called up the Exchange and wrote messages to other children. We then downloaded those messages to an Apple II and then uploaded them to The Source after business hours, when the rates were lower. The same procedure was used for messages going in the opposite direction.

An important observation from this and other projects using long distance communication is that electronic communication tends to be sporadic, and this feature increases with the distance and organizational differences between the sites communicating. It is difficult to create tightly coordinated activities over distances because curricula, schedules, holidays, etc., vary considerably. We ran into this problem continually with the pen pal project. One classroom would become active and send out their messages, while the other class was on vacation or engaged in other language arts activities. By the time the messages were answered, they were old news, or the original children had moved on to other interests. Also, because the messages were being portaged, and in some cases depended on undergraduate assistants, semester breaks also caused delays. So, in spite of the technical speed of the medium, the messages often took much longer to arrive than if regular Postal Service mail had been used! When they did arrive, it was often in bunches. Another cause of sporadic transmission resulted in the relatively sparse amount of communication over the long distance channels. There were not always new messages every day, so that sites logged in to get their messages less frequently. While messages may take only a few seconds to transmit, they may sit for a long time before being responded to.

While pen pal letters may seem like a very simple application of the technology, they actually require a fairly good coordination between sites. If the pen pals were corresponding from home with pen and paper, the situation would be simpler than at school because there is no need to coordinate with what the teacher may see as a priority for the use of the computer and so on. But as we attempted to organize it, the writing activities of one particular child had to be coordinated with another particular child. If one or the other children failed to respond for a few weeks, the activity broke down and led to frustration on both sides.

This exchange of messages became more successful when an energetic UCSD undergraduate took it on as her course project. However, she moved the activity from personal letters to "editorials" about current events. The bombing of Libya took place the day before she began work with the group of San Diego students. The San Diego messages expressing their opinions about the event were answered relatively quickly by the New York students, and several exchanges ensued about this and other events. It is notable that the coordination in this case was facilitated by a dedicated undergraduate on one end and an enthusiastic computer coordinator on the other end. But their efforts were made easier by it being a class project. Any student in either class could write an editorial. The success of the exchange did not depend on a specific student receiving a message from some other specific student on the other side of the continent.

A second project at Bank Street which uses long distance electronic communication is the Mathematics, Science, and Technology Teacher Education (MASTTE) project (Quinsaat, Friel, and McCarthy, 1985; McGinnis, 1986). This has 12 sites around the country which are implementing Bank Street's multimedia science program, *The Voyage of the Mimi*. Nine of these sites have been communicating with us and with each other via a computer conferencing system called Parti which is publicly available through the Source.

The most successful activity that we organized on the MASTTE network illustrates an appropriate level of coordination to attempt to achieve over long distance networks linking classrooms. The "guest expert" series has been very popular among a number of the sites and has elicited a large response. Its structure, however, was much simpler than a pen pal exchange or even the "editorial" exchange because there was no need for any particular child or class to respond to any other child or class. We announced over the network that a series of experts, several of whom were featured in the TV show *Voyage of the Mimi*, would be available to answer questions that are sent in. Teachers got questions from their class and sent them in via the network, or phone in some cases. The expert wrote answers to the questions which were then distributed out over the network. Even this simple activity required

a fair amount of work to coordinate in terms of phone calls to make sure some of the sites remembered the schedule of experts and got in their questions. However, if a site did not respond, it had little effect on the value and enjoyment of the activity by the other sites.

Integrating Local and Long Distance Networks

Where a group of individuals or sites have no intrinsic functional relationship to each other, a considerable amount of extrinsic coordination is necessary in order to use long distance networks for highly coordinated science activities. Networks which have successfully implemented coordinated science activities among distant classrooms, such as the Intercultural Learning Network (Levin, Riel, Miyake, and Cohen, in press) have relied on the supervision by university-based researchers who form a functional research community with purposes beyond the implementation of classroom activities. We also have found in earlier work that a group of district people from around the country could very successfully use the network to plan a conference (Newman and Rehfield, 1985). In this case, the group had a common purpose outside the network activities for which the network was instrumental.

Classrooms in distant cities which join a common network activity often have no connection with one another outside the network activity itself. In these cases, it may be more appropriate to design activities in which sites contribute to a common database of material or information but in which the contribution of any one site is not critical. The guest expert activity meets this description as do "telecourses" in which geographically scattered students take part in "class discussion" using computer-mediated conferencing (Harasim, 1986). Joint data collection activities in which the data from each specific site is critical to the experimental outcome will probably require a considerable amount of extrinsic coordination among distant, unrelated classrooms.

The situation is quite different at the local level of the classroom or school. At this level, planning a science curriculum is not hindered by differences in schedules. Within a classroom, planning meetings are not necessary but even where the activity might involve other teachers in a grade level, or, for example, the school's computer coordinator, face-to-face meetings are relatively easy to arrange. Within the science class, activities can be divided up so different children tackle different aspects then come together for discussions. Using a local area network with a common disk storage device makes that coordination easier and, we hope, more enjoyable for students.

One of the features of the Earth Lab classroom environment is electronic mail. This is a very simple system based on *Bank Street Writer*, the word processor that the schools we are working in are using in writing instruction. We have modified the program to simply add a SEND MAIL and an OPEN MAIL function. In this way, students are able to use their familiar writing environment for sending messages as well. Electronic mail, within the classroom and among the classrooms using a common computer or science lab, simulates the way scientists use electronic mail within a scientific lab or on a single campus. We have devised activities which require groups of students to communicate with others. In this way we integrate science and writing as well as focus students' attention on the importance of clear and unambiguous expression in science.

Following from our experiences with long distance networking, we have also developed an efficient portage connection between the local classroom network and the longer distance networks available via modem connections. Each day the Earth Lab coordinator uses the modem to call out to the Bank Street Exchange, The Source, and other bulletin boards of interest, download messages to the hard disk of the local network and then distribute each message to the directories of the individual students. Ultimately, this function could be automated with a special purpose program.

Through this portage system, we are beginning to solve two of the major problems we have found with networking applications in education. First, by having the children use the same software, in this case *Bank Street Writer*, for all communication—papers, local messages and long distance messages—we will simplify the process by making it necessary to know only one method of entering text. Second, we have reduced the problem of sporadic long distance communications by providing a single source for all communication. If all communication comes over one channel, children are less frustrated by long delays in the long distance networks. Each day, the probability of getting some message is high enough to maintain interest. When the long distance messages do arrive, they simply provide an additional motivator.

Our goal is to make written communication an ordinary and routine part of doing science—*as it is among real scientists*. To achieve this we are protecting children from the difficulties and frustrations of long distance networks, while providing them with interesting communications from outside the school. It is important to simplify the process of communication, otherwise it will continue to be the domain of enthusiasts rather than a routine part of education.

Conclusion

We also want to encourage the development of activities that make use of communication because these will help children to reflect on their own understandings by comparing them to the understandings of the same problem communicated by another. However, our ambitions should be tempered by the realization that the "instantaneous" electronic communication will not overcome the wide differences in the schedules of people in different parts of the country or the world. Non-real time can only be stretched so far. The combination of local and long distance communication into a single system will help to close the gaps in responses. The coordination of local and long distance communication will allow us to make the best use of both. □

References

Harasim, L.M. Computer Learning Networks: Educational Applications of Computer Conferencing. Paper presented at the annual meeting of the American Educational Research Association, San Francisco, 1986.

Levin, J.A., and Riel, M. Educational Electronic Networks: How They Work (and Don't Work). Paper presented at the annual meeting of the American Educational Research Association, Chicago, 1985.

Levin, J.A., Riel, M.M., Miyake, N., and Cohen, M. Education on the Electronic Frontier: Teleapprentices in Globally Distributed Educational Contexts. *Contemporary Educational Psychology*, in press.

McGinnis, M.R. Supporting Science Teachers Through Electronic Networking. Paper presented at the New England Educational Research Organization, Rockport, Maine, 1986.

Mitchell, L.S. *Young Geographers: How They Explore the World and How They Map the World*. New York: Basic Books, 1963 (originally 1934).

Newman, D. Networking Systems for Teacher Education: Design issues. Paper presented at the annual meeting of the Northeastern Educational Research Association, Kerhonkson, New York, 1985.

Newman, D., and Rehfield, K. Using a National Network for Professional Development. Paper presented at the annual meeting of the American Educational Research Association, Chicago, 1985.

Quinsaat, M., Friel, S., and McCarthy, R. Training Issues in the Teaching of Science, Mathematics, and the Use of Technology. In L. Loucks (Chair) *"Voyage of the Mimi": Perspectives of Teacher Education*. Symposium at the annual meeting of the American Educational Research Association, Chicago, 1985.

Slavin, R.E. *Cooperative Learning*. New York: Longman, 1983.

Computer CONFERencing in the English Composition Classroom

Rosemary Kowalski

Computer conferencing is a way in which people who cannot meet face-to-face can communicate easily over an extended period of time using computers. In an educational setting, this form of conferencing is valuable because it enables students and instructors to "talk with" each other beyond the normal classroom's time and space constraints. More importantly, in the composition classroom, computer conferencing promises to be an important new tool.

Computer conferencing takes advantage of a computer network system. Participants can sit at stations on various parts of a campus and "plug in" using appropriate program disks or just signing on if the computer is already directly wired. Participants can even join in many miles from campus if they are properly equipped with a modem connecting their personal computer to the mainframe computer via a telephone hookup. They can communicate with each other at their convenience, which means night-owls can participate till all hours while early risers can work at times more suitable to them. Computer conferencing resembles a telephone in that participants can communicate over long distances, but it differs because the messages are written onto the computer screen and printed out, transmitted and received only when another participant signs on and "catches up" on the discussion. In other words, the discussion is carried on even if some or all participants are not present at once, and participants can retrieve information entered much earlier in the discussion.

The computer conference program I use is Confer II, developed by Advertel Communications Systems, Inc.[1] In this particular program, topics, called "items," are entered and, as the conference continues, the number of items increases, causing the "discussion" to grow longitudinally. However, the discussion also grows vertically as participants respond to each item. Thus, the items increase in number as the participants think of more questions to ask each other, and Item 1, which, for example, might ask an introductory question such as "Where are you from?" grows as each participant answers this question about himself. The program also features "messages," "bulletins," and "notes" enabling other kinds of communications between participants.

I have used Confer for three semesters in five first-year college composition classes. The first term, I introduced it half way through the course to two classes. In one of the sections, which included 23 participants (22 students plus myself), we were on Confer for 8866 minutes. That means that each of us was on for an average of 5 hours and 42 minutes—or just about two extra weeks of class time/discussion. The second section with 22 participants used Confer for 7772 minutes or each five hours and 53 minutes. Last term, 20 participants in my introductory composition class spent 14,326 minutes on Confer over the course of the entire semester, which averages out to an *extra 11 hours and 56 minutes of discussion for each class member*. This almost 12 extra hours of discussion is the equivalent of four extra weeks of discussion if class meets for three hours per week. Clearly this is a significant increase in class interaction. And the types of interaction are significant as well.

Contributions to the Composition Classroom

Expanding Discussion

Confer enables students to explore topics and ideas for writing in ways prohibited in a regular composition classroom. The major reason for this, as pointed out above, is that students devote more time to it. But another reason is that students can get continuing feedback from many more people. The discussion doesn't stop when the hour is up. It is possible to continue discussing a topic for several days, even weeks, providing a richness and depth impossible in ordinary student discourse.

To begin discussion, I encourage students to enter the topics for their next paper into Confer items. I also suggest they try to formulate a thesis and provide some evidence that would back that thesis. I ask the other members of the class to respond, telling the writer whether the thesis and or the evidence seems valid. Some participants are able to suggest other ideas to support a given thesis, a practice which helps those who have difficulty writing longer papers because they cannot think of enough reasons or don't have "any more to say" on a topic. Other respondents point out weaknesses or possible objections to certain arguments. Often such informal beginnings as— "I am going to try to

Rosemary Kowalski is a lecturer in the introductory composition program at the University of Michigan at Ann Arbor.

write my next paper on rock music from the 60s. Do any of you know anything about it? Can you think of some things I might include?"—is enough to get a writer started. Sometimes the writer receives little more than one or two vague suggestions—"My favorite group is the Beatles" or "Be careful! You need to have a lot of examples"—but even these two are more than the writer would have had otherwise working in complete isolation, as is often the case in a regular classroom. More likely, the writer does receive more substantive suggestions.

This kind of interaction is similar to talking over a topic with roommates or friends—a common writing strategy. But the additional benefit of Confer is that the participants share a knowledge of the teacher's instructions (and idiosyncrasies!), the class readings, and in-class discussion, and the conversation includes many more participants and viewpoints than the earlier, pre-computer informal discussion model. Often, on Confer, an "off the cuff" comment one student makes to another enables a third to "see" something she hadn't previously, creating a much more complicated and invigorating discussion. This associative kind of thinking opens up the discussion to ideas which would never have been explored with each student working alone or with just a very limited amount of in-class discussion. Also students can take time to respond. They can formulate their responses at their leisure. They can "pass" and come back to it later if they cannot think of a response immediately. One student remarked that "everyone gets to talk." No one can interrupt another. If a participant dawdles over his answers, the Confer equivalent of an extended vocal "er" or "umm," he will still be able to get his "two cents in."

Students discovered another way of using Confer was to plea for help. One writer entered the introduction to the paper on which he was currently working. He thought the introduction was poor and asked whether the other students had any suggestions. Other students entered sentences they were having problems with on a current writing project and asked for advice. In all these instances, the others quickly provided help which the writers were able to consider and incorporate into their papers even before the rough draft was due in class, a beneficial feature for novice writers.

One student asked "to continue the discussion about Karen's paper where we left off in class on Wednesday." Karen's paper dealt with prejudice and Confer offered us an excellent way to clear up some of the issues and emotions which had been just touched on in class. We were able to pursue this topic more fully and avoided the frustration felt when a class discussion is not finished in time.

Enhancing Social Interaction

While Confer seems to offer the potential for stimulating students intellectually, by far its more obvious contribution is social. For one thing, with just a few items, students can learn a good deal about the other students, from their hometowns, to their majors, to local address, to favorite foods, music, films, etc. They can also provide each other with advice, such as, which instructor NOT to take next term for chemistry—or writing! And they can use the system to discuss the class in a great deal of detail. In short, they function much more like a group of friends—or at least close colleagues—than like a group of people who just happen to be taking the same class. While it is true that other classes, particularly small, intimate writing classes, also create this kind of supportive atmosphere without the aid of a conferencing system, with computer conferencing the class members have the potential to be even closer.

My best example concerns a class I had last term. It was the quietest class I had ever had in all my years of teaching. I tried everything I could think of to get students to relax and speak up and even have a good time. However, nothing worked and, even at the end of the term, the in-class climate was a long way from what I would have liked it to be. But Confer saved us. On Confer, the members did speak. They talked about all manner of subjects. They argued with each other, lent each other support and advice, planned parties, and sent hundreds and hundreds of private messages. They "spoke" on Confer in ways and to an extent they would not in the classroom.

An item later in the term helped us to analyze this phenomenon. One student wrote: "It's funny but when you walk into class everyone . . . . there says absolutely nothing as if we do not know each other at all. On the computer everyone is funny and talkative. T[his] seems like there is something wrong to me!!"—a remark to which I had to heartily agree. But the next student responded: "No. On the computer we can talk about anything we want. It's a break from traditional homework and classes. Perfectly logical . . . ." Still another added: We "feel more bold on Confer." He meant that with no visual or audio feedback—the kind of body language which controls much of what a person says—people feel more free to say what they want.

I have not completely figured out just what caused the muteness in this classroom. I know that several students in the class had been in tutorial classes, and experience had taught me that these students generally are poorer writers and much more reluctant participants in a composition classroom. I also know that some others simply did not want to be in this—or any other—writing class. And

I also know that several of the students were just genuinely shy and would probably not speak up spontaneously under any circumstances. Perhaps, it was just the luck of the draw that all these students just happened to converge in my classroom that semester. While I do not understand the class dynamics exactly, I do know that without Confer, I would have felt completely stymied. In fact, it took me a while to "see" that, in fact, good discussion and interaction was taking place, after all.

Several other instances typify what Confer can contribute to a classroom. One student was miserably homesick during his first semester in college. He entered an item mentioning problems with "adjustment." Again, the responses were quick and warm as classmates tried to offer advice. I think the fact he could raise the issue and talk openly about it helped him to adjust more easily than he might have done otherwise. And it proved to the others that none of them was just a number at a large university; people could and did reach out to others. Another participant sent drawings to the others—a heart for Valentine's Day; a house surrounded by snowflakes after a heavy snowfall. Then, too, there were the frequent printed good wishes—"Happy Snow," "Happy Thanksgiving," etc., which also added a feeling of fellowship to the conference.

One interesting item asked the students their ideas about a Supreme Being. What started out as a possible paper topic, moved quickly to an informal rap session. The student realized his own confusion and soon abandoned the topic. But he was able to see this confusion shared by others, and instead of possibly feeling "stupid," he gained an understanding of the topic's complexity. I don't recall any other classroom discussion on religion being as calm or relaxed yet as open. Similarly, another student proposed writing about "love," a topic which frequently leads first year writers into difficulty. A few responses convinced this writer to abort the attempt and probably saved much time and anguish for the writer and her peer editors.

Other items were just plain fun: (1) "Ask Dr. Larry"—attempted a not very risque imitation of Dr. Ruth. (2) "Hangman"—imitated the children's game now known as "Wheel of Fortune" on television and had the most responses (186) of any item in any conference I've had. (3) "The Great Class Novel"—based on another children's game, allowed each participant to write a piece of the story. (4) "Plans for the Weekend"—notified students where the weekend parties were. (5) "Preparations for class party"—planned a get-together for the students outside of class. It is important to note the non-English topics receive the most re-

sponses. These kinds of items, generated by the students, underpin the good feelings of the conference and the class and are often the place where a sense of fun—and play—originate.

Possible Drawbacks

There are, of course, some concerns about this new tool. One involves time. Including this element in the classroom prohibits using others. I believe all writing classes must provide for some element of play—or a chance for students to explore topics and styles—but whereas in previous semesters, I have always required a journal, I now require participation in Confer. Confer allows students to write informally just as the journal does, but it also allows students to share their work publically with many more people, though they can, if they wish, send private messages to me or to the others in the class.

Another concern about computer conferencing is the suspected anonymity of students using the computer facilities. During the first term I used Confer, I met my class in a computer lab every day. One of my most vocal students entered an item titled, "Memo from Station Three," in which he complained about the "faceless" quality of Confer and the fact that he didn't know who he was talking to when he "spoke." Students began identifying themselves either by naming their station number, or describing what they looked like or where they were seated in the room. And in this way they did get to know each other and began associating responses and actual faces, and the "anonymity" began to disappear. However, another thing I thought significant was that this student felt he could so easily and publically critique the class and by implication me. Normally, a student might complain privately to an instructor or to a few other members of the class, but it is rather rare that this would come up so early in the term and so openly. If anything, this was an indication of just how "relaxed" and interactive the group was. Still more significant were other students' defense of Confer. They confessed their inability to speak out in a classroom, but felt very comfortable and spoke very freely on Confer. For some, this type of public discourse is not nearly as threatening. In the quiet section mentioned earlier, one student only spoke when she was called on. She was very shy and very stiff as she sat in class. Yet she very easily became comfortable with Confer and sent 285 private messages to other members in the class by the end of the term. That resulted in a computer printout of 318 pages measuring 1 and 1/2 inches in thickness for her messages alone. Ironically, what can potentially be a distancing environment—students seated at separate stations staring at computer screens—can become, if conferencing is used, a close, supportive environment. In a regular classroom, where machines are not interferring, Confer helps to increase social interaction even more.[2]

Another point worth mentioning about computer conferencing is that it is more like talking rather than writing. Conversations on Confer flow more like conversations at a party—items "drift" onto new and seemingly unrelated topics and there is not the organization or coherence we expect from a well-written piece. Also, since the writing here is informal, participants make many typing, grammatical, and mechanical errors. Most people who use Confer overlook these sorts of errors, for the conversation and the ideas are more important than formal writing features.

At present, I don't see a good, direct way of using Confer to teach writing as such. I know that others have students enter their papers into the conference and use Confer as a way to publicly critique them, but I am not yet comfortable with this method and I am not yet satisfied that such a use is as good or better than the old-fashioned hard copy/paper method. Perhaps this is simply a matter of taste or style, but the research indicates the difficulty of reading lengthy entries on-screen (Gould and Grischkowsky; Kruk and Muter) and others have suggested simply "plugging in" old paper techniques to the new computer technology has its drawbacks (Strickland). Chief among them is student resistance to what they perceive are simply gimmicks.[3] Consequently, I am exploring the unique qualities of Confer which do not simply mimic old paper and pen methods.

Conclusions

I contend that Confer is not so much a tool for writing practice *per se* as it is a tool for discussion and "play" and since most composition teachers believe in the importance of these elements in their classes, it offers a unique and entertaining method for incorporating them in the classroom. But Confer is more than just a composition toy. The observations I have noted are important ingredients in writing development. As Lil Brannon discusses in "Toward a Theory of Composition," current researchers agree that writing abilities ". . . instead of developing from correctness, to clarity and finally to fluency—the priorities followed in schools . . . develop in precisely the opposite way; from fluency, to clarity, to correctness . . . ."[4] Confer, then, would seem to be a way to gain these skills in the appropriate sequence. And Brannon also cites James Britton's

influential work, which emphasizes the important links between talk and expressive writing and expressive writing and transactional or expository and argumentative writing (18-19). Again, Confer can help bridge these stages in writing. Clearly, computer conferencing contributes to both these important areas of writing development and is a tool which begs for further investigation. □

Notes

1. Confer II program is copyrighted material and a registered trademark of Advertel Communications Systems, Inc., 2067 Ascot, Ann Arbor, MI 48103.
2. At the present time, I have abandoned the computer laboratory classroom because I think I teach to the machines too much; that is, I feel that since they are there I should be using them and this interfers with my other beliefs about teaching composition.
3. In another experiment I did with two of my composition classes, I learned that students resist assignments which they perceive to be merely trying out different ways to use the computer.
4. Interestingly, my first efforts in trying to use computer conferencing as a writing tool backfired because I tried to do last things first. I tried to have students do sentence exercises, correcting sample sentences for practice. And I tried to make them do a formal system of argument analysis following the formula of Jack Meiland in *College Thinking*. However, these efforts did not work. After the first few responses, other students quit contributing suggestions to the sentence exercises, perhaps, because they felt the sentences had been "corrected" already. When students asked for help with "real" sentences, help was readily provided. Similarly, students resisted analyzing other student's topics in a formal way probably because such a system seems to demand a particular answer. The freewheeling kind of association which students enjoy and gladly participate in is prohibited using this method.

References

Brannon, L. Toward a Theory of Composition. In *Perspectives on Research and Scholarship in Composition*. Eds. B.W. McClelland and T.R. Donovan. New York: MLA, 1985, 6-25.

Gould, J., and Grischkowsky, N. Doing the Same Work with Hard Copy and with Cathode-Ray Tube (CRT) Computer Terminals. *Human Factors*, 1984, *26*, 323-337.

Kruk, R.S., and Muter, P. Reading of Continuous Text on Video Screens. *Human Factors*, 1984, *26*, 339-345.

Meiland, J. *College Thinking*. New York: Mentor Books, 1981.

Strickland, J. Computers, Invention, and the Power to Change Student Writing. *Computers and Composition*, 1987, *4*(2), 7-26.

Busy Professionals Go to Class the Modem Way: A New Approach to Distance Learning in the Electronic Classroom

Michael Thombs, Patricia Sails, and Beverly Alcott

"Good afternoon Pat, Don, Carol.... Bev, and Jack, it's good to see that you were able to join us from Germany. We will be starting in just a few minutes. I think we will give the others a few more minutes to join us," says Professor Al Mizell, Director of Nova University's Computer Education Department, the instructor of the class that is about to begin.

Meanwhile, Mike, late to class, is scurrying about trying to find the classroom. "I know they are around here somewhere," he mutters as he begins to lose his electronic cool. "Better ask someone where the curriculum class is being held," he says as he "talks" to another student. "Excuse me, do you know where the curriculum class is being held this afternoon?" he asks. "Not sure, why don't you try classroom one or two?" is the reply. Checking Classroom 1, Mike finally finds his class, with only a minute to spare. "Hi Mike, glad to see you have arrived...we will be starting shortly. While you're waiting, you can jot down some questions you would like to raise in class if you like. Then when I call on you, I'll put them on the blackboard for everyone to see."

This is just one way that many masters, specialist, and doctoral degree candidates go to class at Nova University. So far, you may not be impressed by the classroom scenario described, but perhaps we have left out something. What is it that makes this class so uncommon? To answer that question, let's take a closer look at the participants and instructor:

- Mike almost missed the beginning of class because he was planting some trees in his yard just minutes before entering the classroom.
- Pat was moving that Sunday and paused for exactly sixty minutes to attend class.
- Carol suspended her research with the Library of Congress to attend class and within minutes of class being over, resumed her studies.
- Bev and Jack were approaching their bedtime, having been working on their practicums and research for the past 12 hours (since 9 A.M.); and
- Dr. Mizell was attending a graduation exercise fifteen minutes after the close of class.

By now you probably visualize a classroom of pocket tape recorders all listening to a "reel-to-reel" tape player stationed on the instructor's desk. Well, if none of this makes sense, let me add some logistics to our mystery. Pat went to class in Sarasota, Florida; Carol was at home in New Jersey; Mike was attending the class in Rhode Island; Bev and Jack were participating from their homes in Germany (by European Standard Time it was 9:30 P.M.); and Dr. Mizell was in Fort Lauderdale, Florida. All were in class at the same time, taking notes from Dr. Mizell, asking questions, and interacting with each other. How did they do it? The answer: The Electronic ClassRoom (ECR).

Students enrolled in Nova University's computer-based masters, specialist, and doctoral programs use a wide variety of alternative delivery systems. ECR, designed and developed by Don Joslyn (senior systems programmer at Nova University), is the newest, and by far the most innovative, delivery system used by students, faculty, and staff. It affords students the luxury of attending class and interacting with classmates and their instructor without having to leave home. No need to be distracted by taking notes; ECR is able to make notes of the entire session and save a transcript of the class proceedings for future reference by students. If students miss or need a review of the class, the transcripts are available for "downloading" to their own computers for study at a later time.

How does it work? Students and instructors are connected to a host computer system located at Nova University in Fort Lauderdale, Florida via Tymnet, a commercial long distance telecommunications service. Each student dials into a local node using a standard 300, 1200 and even 2400 Baud MODEM. The host, a VAX 11/780 running UNIX, supports the ECR session.

Earlier we mentioned that Mike was looking for the classroom and indeed he was. The UNIX-based ECR facility supports several concurrent electronic classroom sessions, each capable of supporting a maximum of 24 students. Students looking for their class log into Tymnet, and then into the Nova University system. When students type "ECRS" a list

Michael Thombs is Assistant Professor at Salve Regina College, Newport, Rhode Island. Patricia Sails is a math teacher at New Directions Alternative High School, Sarasota, Florida. Beverly Alcott teaches at The Department of Defense Dependents Schools.

of all ECRs in session appears on their monitors. The students can then reenter the ECR command followed by the identification of the proper class. That is all there is to do to "go to class" when operating within the ECR environment. Although ECR at Nova University is still in its preliminary stages, students now participate in ECR sessions in many masters, specialist, and doctoral classes. ECR, as one of several delivery systems used at Nova University, looks to be a promising addition.

How are the students able to interact while logged into the UNIX environment of Nova University? To provide the reader with some background, the UNIX environment supports a "talk" command (TALK is a UNIX-based online interpersonal communications facility) that gives two users the ability to converse online. In seeking information about the location of his class, Mike first took a look at "who" was currently logged onto Nova's computer. Using the "talk" command, he was able to ask for the location of his class from another student he "met" in the UNIX electronic hallway. The talk facility divides the screen into two sections. While one user types messages on the top half, the other user replies by typing messages on the bottom.

ECR behaves in a similar fashion. The screen is divided into two parts as in the talk facility. The top two-thirds of the screen is called the blackboard. The instructor controls the blackboard, using it to lecture, provide instructions or information, and answer student questions. The instructor can also pass control of the blackboard to any student in the Electronic ClassRoom. The bottom portion is used by the students participating in the ECR. In this area, the UNIX-based ECR facility lists each student's identification names (standardized as last name and first initial). The last four lines at the bottom are reserved for message composition.

Prior to a class session, the instructor is able to prepare several text frames, much like overhead transparencies or lecture notes. These might be questions and answers, charts, lists, and/or short case studies. The frames or series of frames are stored as standard UNIX files. The instructor may then use these frames to conduct the class, pausing for questions and interactions from students. The instructor can tell when a student in the class wishes to speak by checking a dynamic class list (the list is updated every time a new user enters or leaves the ECR session) of participants, displayed on the bottom portion of the screen. For example, if the instructor wishes to ascertain that everyone is on task and ready to begin or continue, s/he will ask, "Everyone ready?" At this prompt, class members press the "y" key on their computer keyboards and the instructor sees a "YES" indicator beside the name of each member. Students show their desire to "speak" (question, comment, or interact with the instructor and students) by pressing the "a" keys on their own keyboards. The instructor and all the other students can see a question mark next to each student's identification name. The instructor may then turn temporary control over to that student.

As you might guess, the one drawback to an ECR is speed. Since participants often type slowly and only one student at a time can "talk," this can be a source of frustration. To ease this problem, author Don Joslyn has given users the ability to compose and edit short questions, comments, and replies without leaving the ECR or interfering with the class. When called on later, the student can transmit the "prepared" message, question, or response, so that the display appears almost instantly on everyone's terminal screen.

ECR is new to all doctoral students attending Nova. While ECR is still novel, it will be difficult to evaluate objectively the effectiveness of the delivery system until more professors use it to conduct their classes. In addition to the pedagogical questions this medium raises, the ECR also presents many technological problems as would be expected with any new technologically-based system. Participants in our first ECR session used Apples, IBMs and compatibles, Tandys, and many others located in areas all across the country and the world. We then add to this mixture a host computer system, itself based on many more layers of complex software. A reasonable analogy would be to think of a class of foreign exchange students trying to learn algebra each in his own language (protocol) and using his own textbook (terminal emulation and configuration). Finally, added to these problems is the students' unfamiliarity with telecommunications. Many have had little previous communications exposure. Terminal emulation and protocol are words that soon become familiar to all who participate in Nova's online computer-based learning program.

With the help of ECR, all the students need is the will to survive, a good manual, a sense of humor, and many hours of unrestrained creativity. When all these elements come together, great things begin to happen, as we are now discovering. □

AIDS Training in Third-World Countries: An Evaluation of Telecommunications Technology

Pamela D. Hartigan and Ronald K. St. John

Introduction

Scientific conferences traditionally attract a select group of specialists in focused fields of study who gather to debate and deliberate about particular medical or social science issues. Their only audience comprises colleagues within the immediate conference room, and dissemination of their deliberations appears in summary papers or documents.

With the emergence of the human immunodeficiency virus (HIV), the need to reach simultaneously a much wider audience of health care workers has become crucial. During the last five years, the response has been the International Scientific Conference on AIDS. The most recent one, held in Montreal, Canada in June, 1989, was attended by 12,000 to 15,000 people at a cost of approximately 30 million dollars (estimated cost per person, $2,000). The need to disseminate technical information is particularly great in Latin America, where the velocity of the spread of the epidemic in certain countries is greater than in North America, where, to date, most cases have been registered.

Since 1983 the Pan American Health Organization (PAHO), Regional Office of the World Health Organization (WHO) in the Americas, has maintained hemisphere-wide surveillance for infection by HIV. As of June 15, 1989, a total of 115,000 cases of AIDS has been reported by PAHO by 45 of the 46 nations and territories in the Western Hemisphere (Pan American Health Organization, Regional AIDS Prevention and Control Program). Only one territory, Montserrat, has yet to report a case of AIDS or a person infected by the AIDS virus.

Pamela D. Hartigan, PhD., and Ronald K. St. John, M.D., M.P.H., are with the Pan American Health Organization, part of the World Health Organization, 525 Twenty-Third Street, N.W., Washington D.C. 20037. Dr. Hartigan is a behavioral psychologist. Dr. St. John is Director of the Health Situation and Trend Assessment Program of the Pan American Health Organization.

In response to the epidemic, all countries and territories in the Western Hemisphere have developed National AIDS Prevention and Control Programs, following the guidelines established by the PAHO/WHO Global Program on AIDS (1987, World Health Organization). It is clear now that in order to deal with the scope and complexity of the epidemic, all sectors of the national and international community must respond. Medical personnel and health care workers as well as decision-makers, social scientists, and other leaders must participate in these National Programs within their respective countries. It is they who must decide on policies as well as implement and monitor the necessary health interventions to stop the transmission of the virus. These individuals must also coordinate and support biomedical, epidemiological, social, behavioral, and operational research, in addition to developing surveillance, forecasting, and impact assessments.

The applicability of teleconferencing to a broad range of areas, from business and marketing to medicine and art has been documented (Azarmsa, 1987; Ratcliff, 1985; Thiel, 1984). Because this method combines timeliness and immediacy of information with inservice training and cost-effectiveness, teleconferencing has expanded the horizons of education (Azarmsa, 1987). Recognizing the broad array of health care personnel that are critical for the fight against AIDS, PAHO/WHO considers the use of telecommunications technology as an effective and cost-efficient method to provide rapidly the latest technical information to the many different types of health care professionals involved in the National Programs. Bringing many thousands of health care workers together as in Stockholm (1988) and Montreal (1989) to learn about AIDS has many limitations for participation by persons from developing countries due to the relatively high cost of attendance.

First Teleconference on AIDS

In 1987, PAHO/WHO organized the First Pan American Teleconference on AIDS, which was broadcast from Quito, Ecuador, to over 40,000 health care workers at 300 reception sites throughout the Americas. PAHO had decided to bring the most recent research and programmatic information on AIDS directly to the health care worker, via satellite television. Ninety-seven percent of those Teleconference viewers polled felt that more teleconferences should be organized to provide necessary information on all health issues related to AIDS (Pan American Health Organization, 1987). In response to the demand for updated technical information, PAHO/WHO organized a Second Pan American AIDS Teleconference, which was broadcast from Rio de Janeiro, Brazil, on

December 12-14, and reached 36 countries and territories including seven sites outside the Western Hemisphere. A rigorous evaluation methodology was designed and implemented to assess the results of this Teleconference. This article presents the findings of this evaluation.

Evaluation Methods

The Teleconference was an educational effort to disseminate crucial information on AIDS to a large target audience composed of medical personnel and health care workers, as well as decision-makers and social scientists who must deal with the AIDS pandemic in the countries of the Americas. This target audience was comprised largely of individuals who normally would not be able to attend an international conference because of limits placed by budgetary constraints in their developing country settings. Costs of international air travel, conference registration, and hotel accommodations act to exclude the vast majority of these professionals.

PAHO developed an eighteen-hour scientific program which was transmitted by television in four languages over a three-day period. The program consisted of three plenary sessions, four panel discussions and five question-and-answer sessions in addition to special presentations by dignitaries and three documentaries specifically produced for the Teleconference. Each of the five question-and-answer periods screened an average of 200 to 300 questions from participating sites throughout the Region, for a total of 1,000 to 1,200 questions for the entire conference.

There were four purposes to the Teleconference evaluation: to determine the size of the target audience; to identify the composition of that audience; to determine the participants' reactions to each of the eleven presentations; and to determine the efficacy of the forms of presentation, i.e., documentaries, plenary sessions, panel discussions, and question-and-answer sessions.

Twenty-four countries were selected for participation in the evaluation of the AIDS Teleconference. Evaluation booklets were specially designed and distributed to main reception sites only. The sample chosen was a convenience sample, drawn from those participants at Teleconference sites. Upon registration at the Teleconference site, participants were informed that their particular site had been selected to evaluate the Teleconference, and individuals were asked to help in this endeavor. Each volunteer who accepted was given the evaluation booklet containing an introductory page with instructions for completing the evaluation, a page requesting demographic information on the evaluator, and eleven subsequent pages corresponding to the program components and their respective question-and-answer sessions.

Participants were requested to evaluate each session as soon as possible after it was completed. Each page was perforated so as to allow the evaluator to tear off the completed page and deposit it in a box designated by the survey coordinator at each site. An identification number at the top of each page created a link between the demographic information and responses to each session.

Results

All results were analyzed from a country-by-country as well as a regional perspective. For purposes of this article, however, only the regional findings will be presented.

Size of target audience: A total of 41,000 persons attended the Teleconference at multiple designated sites. In some countries, such as Brazil and the Dominican Republic, the program was transmitted simultaneously to designated sites and to national television networks for total coverage of the population. Based on reports received from such networks in various countries, approximately 50,000 to 75,000 additional persons viewed the program.

Size of Evaluating Sample: A total of 3,100 evaluations were distributed to the twenty-four designated main sites. Out of this total, approximately 1,600 were returned. However, about 25% of these did not include the demographic information included with the evaluation of sessions which would allow a link to be made between evaluator responses of the program components and demographic information. For purposes of this assessment, a total of 1,200 responses were used, approximately 40% of the total distributed throughout the Region.

Audience composition: The sample of 1,220 participants who evaluated the Teleconference throughout the Region of the Americas was stratified into three groups: physicians/non-physicians; persons who do/do not work with seropositive individuals; and persons working in their current occupation for 1 to 4 years, 5 to 9 years, and 10 + years. Of the total sample, 55% were women. Among those who indicated area of work, 78% were in the public sector, 21% were in the private sector, and 1% indicated working in both areas.

Physicians made up 45% of the sample. Almost half (48%) of this group were involved in general/family medicine. Of the remaining, 14% were in pediatrics, 12% in internal medicine, 8% in gynecology, 7% in public health, 6% in laboratory science, 3% in psychiatry, and 2% in surgery.

Non-physicians comprised 55% of the sample, and the majority (56%) were involved in health care. The remainder included laboratory technicians, students, and researchers as well as administrators, public information professionals and social workers.

The majority of the evaluators (72%) reported working in AIDS-related activities. Of this majority, one-fourth worked in the area of medical care, while one-fifth worked in health education/promotion. Of the remainder, 12% worked in laboratory research, 10% in surveillance and epidemiology, 5% in field research and social services, respectively, and 4% in patient counselling, public information, and program management, respectively. The remaining 8% were spread across other categories. Approximately half the sample of this group reported working directly with seropositive individuals.

Participant Reactions: Five variables were chosen to evaluate the data: complexity, interest, relevance in relation to current occupation, length of presentation, and knowledge acquired. After initial examination of the data, these five variables were collapsed into three. This decision was based on the high dependence of scores for *interest* and those for *relevance* in relation to individuals' professional backgrounds. Likewise, *length* of presentation was also highly related to *interest*, and those who felt the presentation was of average to low interest also found the presentation too long. *Length* was also related to *complexity.* The more complex the material was rated, the higher the tendency to rate the session as too long. The analysis focused on three variables which appeared independent from one another: complexity, interest and knowledge acquired.

Thirty-five to fourty-five percent of evaluators rated the documentaries as simple in *complexity,* whereas 45-55% rated the plenary presentations as average in complexity, save for the plenary presentation on Retroviruses and Immunopathogenesis, which evaluators (44%) rated as too complex in relation to their professional backgrounds. The panel discussions on Perinatal and Sexual Transmission were rated as most *interesting* of all the sessions by 80% of the evaluators, while 60 - 75% rated the rest of the sessions as very interesting. Results for *knowledge acquisition* indicated that between 80-90% reported learning new information from both plenary sessions and panel presentations.

When the data were examined by profession, both physicians and non-physicians rated the presentation on Retroviruses and Immunopathogenesis as too complex given their professional backgrounds.

With respect to format, the majority of non-physicians reported being more *interested* in panel discussions, giving considerably higher ratings to the panels on Sexual, Perinatal, and Blood Transmission than to any of the plenary or documentary presentations. By comparison, physicians appeared less affected by format and tended to focus on session content. For example, physicians gave similarly high ratings for interest to the plenary on Natural History and Clinical Aspects, and the panels on Sexual and Perinatal Transmission. However, the majority (90%) of physicians reported *learning* more from the plenary format than from any other format. In contrast, non-physicians reported learning equally through both plenary and panel formats.

In summary, whereas physicians focused on special session content when asked about *interest,* they reported *learning more knowledge* through plenary presentations, even where they rated the presentation as too *complex* in relation to their professional background. Physicians, therefore, may, by virtue of training, feel they learn more in a lecture mode in comparison to a discussion mode. In contrast, non-physicians were more *interested* in the panel discussion format and reported *learning* equally through both formats.

Discussion

The data provided by this evaluation were obtained from a convenience sample. In addition, there was no attempt to insure specific sample size by strata. Therefore, in cases where responses related to the participants' profession or the length of years in a current profession are compared, resulting percentages which vary by as much as 10% may not indicate a substantial difference.

Based on the findings of this evaluation, the Second Pan American Teleconference on AIDS reached a large audience of health care workers, the majority of whom were already working in AIDS-related activities. The majority of participants reported learning new knowledge through all four modes of presentation, i.e. documentaries, plenary presentations, panel discussions, and question-and-answer sessions.

The category of physician/non-physician appeared to be the best at discriminating the differences between and within groups for preference of sessions and formats, and results show that this former group report learning the most through the plenary mode of presentation. In addition, reported knowledge acquisition is independent of level of reported complexity, and many physicians reported learning the most from the session on Retroviruses and Immunopathogenesis, although this session was rated as too complex in relation to their professional background. The sessions receiving the highest ratings for level of interest were the panels on Sexual and Perinatal Transmission. The results of the evaluation will permit future refinements in the content and presentation of Teleconferences designed to provide updated information on AIDS.

Conclusion

The overall cost of the Teleconference in Rio de Janeiro was approximately 2.3 million dollars including in-kind contributions. It reached an estimated audience of 40,000 to 115,000 health care workers, at an estimated cost of $20-$58 per capita. In general, the cost for a Latin American or Caribbean professional to attend an international conference in another country is approximately $2,500 to $3,000. Because a large number of persons who participated in the Teleconference will never attend an international conference, PAHO concludes that the Teleconference by satellite is an effective, cost-efficient way to reach a large audience to mobilize human resources for disease prevention and control. □

References

Azarmsa, R. Teleconferencing: An Instructional Tool. *Educational Technology*, December 1987, *27* (12), 28-29.

The Pan American Health Organization. *AIDS Surveillance in the Americas*. PAHO, Health Situation and Trend Assessment Program, June 15, 1989.

The Pan American Health Organization. *Evaluation of the First Teleconference on AIDS*. PAHO, Health Situation and Trend Assessment Program, December, 1987.

Ratcliff, J.L. Statewide Teleconference Educational Programming: Strategic Planning Issues. *Catalyst*, 1985, *14* (1), 12-16.

Thiel, C.T. Teleconferencing: A New Medium For Your Message. *InfoSystems*, April 1984, 64-68.

World Health Organization. *Guidelines for the development of a National AIDS Prevention and Control Programme*. WHO AIDS Series, Geneva, 1987.

Inservice Training Via Telecommunications: Out of the Workshop and Into the Classroom

Joseph J. Stowitschek, Brent Mangus, and Sarah Rule

Introduction

Inservice training with teachers has historically been conducted in a workshop type format. Inservice training carried out through various telecommunications procedures, however, can allow more contact between consultant and teachers, thus increasing the benefits to teachers and children. The employment of interactive audio-video telecommunications allows consulting to occur in the classroom at school sites. Enhancing audio-video telecommunications through the use of electronic mail and audio teleconferencing can further increase a consultant's working time with a classroom teacher. When all three modes of telecommunications are properly applied, the on-site training by a consultant becomes an efficient way to address a variety of training priorities.

Telecommunications Advances

Considering the advances made in telecommunications technology and its availability for educational uses, applications of the technology have been surprisingly restricted. One of the primary advantages of this technology is that it permits interactive communication between a number of people who are not in the same place at the same time. This communication may be similar to face to face, allowing individuals to see and hear each other; it may be limited to audio-only conferences between more than two people; or it may be by electronic mail. To date, two-way telecommunications in education has been applied mainly to high school and college course instruction, workshop activities, and business meetings (Johnson and Amundsen, 1983; Polcyn, 1981). While the technology bridged the distance gap between participants, the content of the applications was similar to content that would have been addressed in face-to-face meetings.

Much of the interactive programming done with telecommunications is at a relatively primitive level. Many applications already discovered in the behavioral sciences have yet to be applied (Fitzpatrick and Beavers, 1978; Daugherty and Mertens, 1978; Polcyn, 1981; Filep, 1980). For instance, there has not been widespread application of telecommunications to train or assist teachers and school administrators with the administration of programs, assessment of children for placement or programming, or to help school systems in need of specialized support services.

Granted, verbal communication similar to that which occurs in meetings and conferences has been greatly enhanced by interactive telecommunications. The reach of higher education courseware has certainly been extended. There has even been some enhancement of teacher-student interactions through the use of remote question-answer sessions (Showalter, 1982). However, the interactive potential of telecommunications has hardly been tapped. When the telecommunications educator relies primarily on expository teaching modes (e.g., "talking heads," graphic illustrations, etc.) the interactive potential of telecommunications systems is used to little advantage.

One application of telecommunications technology in inservice teacher training is to use interactive capabilities to allow consultants actually to observe what happens while teachers are teaching in classrooms, and to train teachers based on those observations (cf. Cavallaro, Stowitschek, George, and Stowitschek, 1980). Normally, conducting direct classroom observations and providing immediate supervisory feedback to teachers requires considerable time and travel. Through the use of telecommunications, most of the consultant's time can be focused on the classroom instead of on traveling to or between classrooms. The content of consultations can thus be focused on helping the teacher learn teaching skills and enhance the application of those skills. The teacher training can be done on the job, requiring little or no absence from the classroom.

Using telecommunications to shape teaching *in situ* can result in immediate benefits for both teacher and students. Through this mode teachers are able to receive immediate feedback while they

Joseph J. Stowitschek and Sarah Rule are with the Developmental Center for Handicapped Persons, Utah State University, Logan, Utah. Brent Mangus is a faculty member at the University of Nevada at Las Vegas. The authors express their appreciation to the staff of Project Passage Preschool and Southern Utah State College who participated in the project, particularly to Virginia Higbee and Nancy Glomb for their assistance. This project was supported by the U.S. Department of Education; no official endorsement should be inferred.

are teaching. Such feedback, albeit administered more directly, has been demonstrated to be effective in helping teachers learn specific skills (Ford, 1984; Cossairt, Hall, and Hopkins, 1973). It is clear from this research that teacher behavior changes over an extended period of time, as repeated alternations between feedback and teaching practice episodes occur (Rule, 1972). The use of feedback by telecommunications may have particular applications where teachers encounter significant learning problems, such as when handicapped children are mainstreamed. These children often require more and specialized teacher assistance to learn to attend to tasks and acquire other appropriate behaviors to help them function in the regular classroom.

The purpose of this article is to demonstrate one application of interactive telecommunications technology designed to provide onsite training to teachers. The focus of this description is on training procedures rather than the hardware in order to illustrate the complexities involved in developing telecommunications-based training technology.

Interactive telecommunications were used to provide consultation and supportive supervision in a mainstream preschool program. The telecommunications process consisted of two-way interactive television made possible by a microwave telecommunications system connecting the University of Utah campus in Salt Lake City with Southern Utah State College (SUSC) in Cedar City. To transmit the signal between the preschool and the SUSC campus in Cedar City, a laser link was used so that consulting personnel were able to view and hear the teachers and students in their classroom approximately 275 miles away. At the same time the teachers could view and hear the consultants without leaving the classroom.

In addition to using two-way audio-video communications, the consultants took advantage of electronic mail and audio-teleconferencing. Initially, when a specific teleconference was planned, the consultants typed a written plan or protocol into the Special Net bulletin board to be transmitted to the teacher at the remote site. Teachers then made changes in this protocol as necessary and returned those changes to the consultants via Special Net. The resulting agenda was based upon the needs of the teacher and student at the rural school. This type of advance planning preceded each session and addressed the implementation of students' individual education program (IEP) objectives and teaching techniques used by the teachers. Also, timelines to be followed during upcoming teleconferences were delineated during the advance planning. In some instances, audio teleconferences between consultants and teachers were conducted to facilitate communication on planning. This advance planning occurred prior to each session in a series of teleconference sessions aimed at improving student learning through enhancement of the teacher's effectiveness.

A Telecommunications-Based Training Program

Training in four teaching formats was provided as the telecommunications system was formatively evaluated: (1) group behavior management; (2) intensive one-to-one tutorial instruction; (3) coincidental teaching during naturally occurring incidents; and (4) group-circle instruction. The group behavior management techniques were focused on increasing children's attending to the teachers. The one-to-one tutorial technique, termed microsession training, involved intensive instruction focused on a single, short-term objective. For incidental teaching, participants were trained to intersperse planned teaching incidents into ongoing activities (termed coincidental teaching). These incidents pertained to preschoolers' objectives on naming colors and labeling or describing objects. In the group-circle format, *Let's Be Social* (Killoran, Rule, Stowitschek, Innocenti, Striefel, and Boswell, 1982) a 26-unit curriculum developed to teach social skills, was used by participating teachers.

Telecommunications consultations on the four teaching formats overlapped, but there was a major focus on one format during successive sessions. The overall teleconference schedule is described in Table 1 followed by a more detailed description of consulting activities carried out regarding one of the formats—microsession training.

The focus of conferencing for each week was as follows:

- Week 1— Audio Conferencing to establish schedule of two-way transmissions, child objectives, and teaching tactics to be implemented.
- Week 2— Introducing teachers to the teleconference system and the capabilities of the specific system being used. Introducing students and teachers to the live two-way audio-video process (Table 1, Week No. 2).
- Week 3— Assisting the classroom teacher with positive reinforcement techniques and methods of increasing time on task of the handicapped students.
- Week 4— Demonstrating the microsession teaching techniques by consultants. Teachers were instructed

Table 1

Telecommunication Schedule for a Twelve-Week Period

Week	Telecommunication Mode	Training Topic(s)
1	Audio teleconferencing Electronic Mail	Introduction to materials (10 min.) Administrative questions (5 min.) Microsession (15 min.)
2	Audio-Video Microwave Transmission Electronic Mail	Demonstrations Microsessions (30 min.)
3	Audio-Video Microwave Transmission	Behavior Management (10 min.) Prescriptions Microsession (20 min.)
4	Audio-video Microwave Transmission Electronic Mail	Demonstrations (10 min.) Microsession (20 min.)
5	Audio-video Microwave Transmission Electronic Mail	Microsession (30 min.) Coincidental teaching (10 min.)
6	Audio-video Microwave Transmission Electronic Mail	Microsession (20 min.) Coincidental teaching (20 min.)
7	Audio-video Microwave Transmission Electronic Mail	Microsession (10 min.) Coincidental teaching (30 min.)
8	Audio-video Microwave Transmission Electronic Mail	Microsession (10 min.) Coincidental teaching (30 min.)
9	Audio-video Microwave Transmission Electronic Mail	Microsession (10 min.) Coincidental teaching (20 min.) *Let's Be Social* (10 min.)
10	Audio-video Microwave Transmission Electronic Mail	Microsession (10 min.) Coincidental teaching (10 min.) *Let's Be Social* (20 min.)
11	Audio-video Microwave Transmission Electronic Mail	Microsession (10 min.) Coincidental teaching (10 min.) *Let's Be Social* (20 min.)
12	Audio-video Microwave Transmission Electronic Mail	Microsession (10 min.) Coincidental teaching (10 min.) *Let's Be Social* (20 min.)

step-by-step through the micro-session sequence on teaching to a specific IEP objective with a student.

- Week 5-6— Refining teacher's use of the microsession teaching process. The advance planning via electronic mail for subsequent sessions included a complete description of a new teaching technique termed coincidental teaching. The teachers were asked to read and discuss the coincidental teaching methodology and to attempt to use this new method one time before the next teleconference.

- Week 7— Observing and demonstrating coincidental teaching. The teachers in the school attempted to apply coincidental teaching on request as the consultants observed using the two-way system. The consultants then provided a demonstration to clarify the coincidental teaching methodology.

- Week 8-10— Demonstrating and refining microsessions and coincidental teaching. The teachers continued to demonstrate during each teleconference a short example of the microsession and coincidental teaching activities they were currently using with students. Also, the teachers requested assistance to help one student become more socially responsive. The consultants mailed a *let's Be Social* curriculum to the teachers. The teachers examined *Let's Be Social* and chose three units that they felt could benefit the student as well as the entire class.

- Week 11— Training in *Let's Be Social*. The teachers implemented the first unit they had chosen from the *Let's Be Social* curriculum while being observed by the consultants. The consultants then provided feedback that was directed at refining the teaching methods used.

Figure 1

*Pattern of Telecommunications
Consulting Activity*

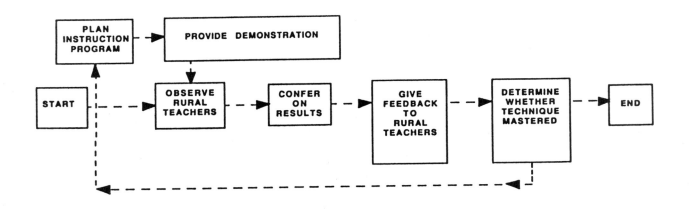

- Week 12— Continued refinement of the *Let's Be Social* teaching methodology.

Throughout the latter training sessions, the teacher's use of microsession and coincidental teaching techniques continued to be demonstrated on the microwave system. The consultants monitored these programs to ensure that teaching behaviors and child progress were maintained but expended the greatest amount of effort on the *Let's Be Social* curriculum implementation. This approach was consistent throughout the entire sequence of sessions.

Telecommunications Session

Prior to any audio-video microwave transmission, the day's schedule was sent to trainees through electronic mail. The sample session described below followed previous sessions in which the teaching program and procedures had been specified and trainees observed while teaching. A standard pattern of procedures was followed as shown in Figure 1.

Microsession Teaching

1:30 p.m. The consultants began the teleconference by stressing the important points of the upcoming teaching demonstration. A two or three phase cue card included key phrases to assist the teacher to remember the points to be addressed during both the consultant's demonstration and the subsequent teaching practice session (See Figure 2).

1:40 p.m. A teacher practiced the teaching technique as demonstrated by the consultants. The practice session was with the target student on an IEP objective previously outlined. The typical practice session was three to five minues in duration. Consultants observed the teacher and recorded the results on a data sheet (see Figure 3).

1:45 p.m. The consultants, having observed the teacher implementing the new technique, conferred with each other and provided feedback as to specific deficiencies and positive points of the teacher's techniques. Consultants also made recommendations for changes in teaching strategy that may assist both teacher and student.

Consultant feedback was based on a predetermined set of teaching criteria outlined by the consultants. The verbal feedback to the teacher was provided by a single spokesperson who compiled information from the consultants and relayed it to the teacher.

1:49 p.m. The teacher was requested to repeat the teaching sequence following the consultant's suggestions. The consultants observed the teacher to determine if their feedback and/or recommendations for change had helped the teacher.

1:54 p.m. Again the consultants reviewed the data sheets to determine the defi-

Figure 2

Example of a Cue Card Used
to Prompt Participating Teachers

1. <u>Get child's attention</u> .

2. <u>Correct errors</u> .

3. <u>"Ham it up"</u> .

Figure 3

Portion of the Observation Form Used
During the Microwave Teleconference

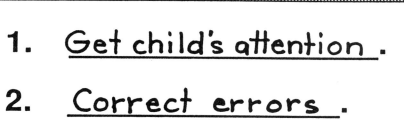

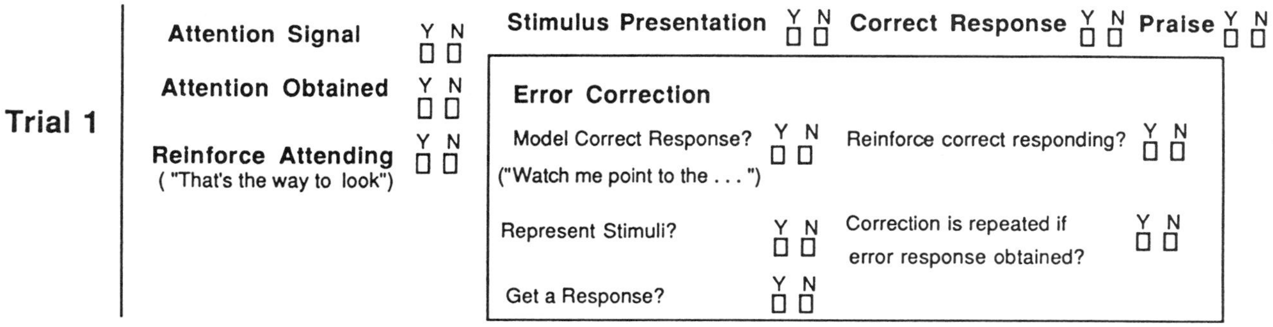

Trial 1

Attention Signal Y N ☐ ☐

Attention Obtained Y N ☐ ☐

Reinforce Attending Y N ☐ ☐
("That's the way to look")

Stimulus Presentation Y N ☐ ☐   **Correct Response** Y N ☐ ☐   **Praise** Y N ☐ ☐

Error Correction

Model Correct Response? Y N ☐ ☐   Reinforce correct responding? Y N ☐ ☐
("Watch me point to the . . . ")

Represent Stimuli? Y N ☐ ☐   Correction is repeated if error response obtained? Y N ☐ ☐

Get a Response? Y N ☐ ☐

---

cient and superior points of the teacher's teaching method. The consultants provided feedback through the spokesperson and then provided a time when there was open communication so that the teacher could ask questions about the teaching technique or the instruction program.

2:00 p.m. Conclusion of the teleconference. Follow-up teaching activities between this and the next conference were outlined and agreed upon. See Figure 4 for a view of the entire process.

Discussion and Conclusion

Research on the effectiveness of feedback provided using the microwave system is currently underway. However, a formative evaluation was completed in which a consultant observed two participating teachers as they attempted to use prescribed teaching procedures. Both teachers demonstrated increased accuracy in their use of

Figure 4

Illustration of Four Major Consultation Activities

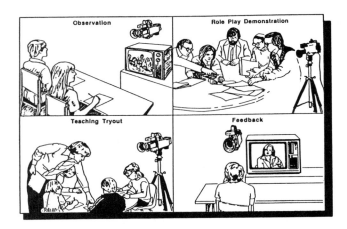

the procedures as they received training through the microwave system.

The use of telecommunications in consultations on classroom teaching offers several benefits. Travel time can be greatly reduced, resulting in a savings not only in time but in money. Repeated contacts with teachers can be made over time, rather than relying on a single workshop or consultation visit. The quality of the service provided to the school teacher can be high, and the service can be provided to an increased number of teachers due to the savings incurred. While the interactive telecommunications system does not replace the personal contact required to make a consultant and teachers comfortable with one another, it can increase the amount of professional contact and thus the benefits to the personnel involved. □

References

Cavallaro, C., Stowitschek, C., George, M., and Stowitschek, J. Intensive Inservice Teacher Education and Concomitant Changes in Handicapped Learners. *Teacher Education and Special Education*, 1980, *3*(3), 49-57.

Cossairt, A., Hall, R.V., and Hopkins, B.L. The Effects of Experimenter's Instructions, Feedback and Praise on Teacher Praise and Student Attending Behavior. *Journal of Applied Behavior Analysis*, 1973, *6*, 89-100.

Daugherty, D., and Mertens, D.M. *A Summative Evaluation of Teaching the Group Handicapped Child.* Technical report prepared for the Department of Health, Education, and Welfare, Washington, D.C. (ERIC Document Reproduction Service No. Ed 165 810, 1978).

Filep, R.T. *Telecommunications and The Rural American, Today and Tomorrow.* Background materials for a workshop on telecommunications in the service of rural education. Washington, D.C. National Institute of Education (ERIC Document Reproduction Service No. ED 229-203, 1980).

Fitzpatrick, J.L., and Beavers, A. *An Inservice Course on Mainstreaming: An Innovative Approach.* Toronto, Canada. Paper presented at the annual meeting of the American Educational Research Association (ERIC Document Reproduction Service No. ED 154 556, 1978).

Ford, J.E. A Comparison of Three Feedback Procedures for Improving Teaching Skills. *Journal of Organizational Behavior Management*, 1984, *6*, 65-77.

Johnson, M.K., and Amundsen, C. Distance Education: A Unique Blend of Technology and Pedagogy to Train Future Special Educators. *Journal of Special Education Technology*, 1983, *6*, 34-45.

Killoran, J., Rule, S., Stowitschek, J.J., Innocenti, M., Striefel, S., and Boswell, C. *Let's Be Social.* Logan, UT: Outreach, Development, and Dissemination Division, Developmental Center for Handicapped Persons, Utah State University, 1982.

Polcyn, K.A. The Role of Communication Satellites in Education and Training: The 1990's. *Programmed Learning and Educational Technology*, 1981, *18*, 230-244.

Rule, S. A Comparison of Three Different Types of Feedback on Teacher's Performance. In G. Semb (Ed.), *Behavior Analysis and Education.* Lawrence, Kansas: University of Kansas, Support and Development Center for Follow-Through, 1972, 278-289.

Showalter, R.G. *Speaker-Telephone Continuing Education for School Personnel Serving Handicapped Children.* Final project report to Indiana Department of Public Instruction, Division of Special Education (ERIC Document Reproduction Service No. ED 231 150, 1982).

Suarez, T.M., Clifford, R.M., Montgomery, D., Rodrigues, E., Kuligowski, B., and English, J.J. *Telecommunications Training for Day Care Teachers, Aides, and Administrators: A Comparative Study.* Frank Porter Graham Child Development Center, University of North Carolina at Chapel Hill, 1984.

Extending Education Using Video: Lessons Learned

Dean R. Spitzer, Jeanne Bauwens,
and Sue Quast

Introduction

Increasingly universities are being forced to experiment with alternative delivery systems for providing quality education to growing populations of undergraduate students. This is especially true at state universities with a mandate to admit all qualified students. At Boise State University, this is precisely the situation. Student numbers have grown to a point at which the use of alternative delivery systems is essential.

In the College of Education, where enrollment is growing much faster than faculty, there is a strong challenge to innovate. There are many options for extending the ability of faculty members to reach greater numbers of students. Research has clearly demonstrated that no one delivery system is ideal in all situations (Schramm, 1977). Depending on the situation, video, computers, and other media formats have been shown to be effective for some purposes (especially when they have been used well).

At Boise State University, video is a particularly attractive medium to help us to cope with increasing student demand, to extend educational opportunities to distant learners and to cope with large class sizes in a sub-optimally funded state university system. We have an abundance of video resources, especially through the new six-million-dollar Simplot-Micron Technology Center (funded primarily through private donations), which provides the campus and the community with outstanding media resources.

The Study

An evaluative research study was designed to investigate the feasibility of using video-based in-

Dean R. Spitzer is Associate Professor of Education, Program Director of Instructional Technology, and a Contributing Editor to this magazine, **Jeanne Bauwens** is Assistant Professor of Education, and **Sue Quast** is Coordinator of Instructional Television, all at Boise State University, Boise, Idaho.

struction (supplemented by graduate assistant monitors) for high-demand required courses. It is hypothesized that such a format will enable the College of Education to gain the most productive use of faculty resources and the most appropriate use of graduate assistants. We wanted to see how the students would respond to this new (for Boise State at least) learning format in terms of course achievement and attitudes toward various aspects of the delivery system.

This study was composed of three groups: two comparison groups and one that participated in the course via videotape. The video group received the eight-week one-credit course titled "Education of Exceptional Secondary Age Children," via videotape monitored by a graduate assistant. One comparison group participated in the course at the Simplot-Micron Technology Center, as the course was being videotaped. The other comparison group participated in the course in the Education Building in a completely conventional lecture environment. The course covered the identical content for each group and is required for all students seeking secondary teaching certification. The same instructor taught all groups. At the end of the course, in addition to the course exam, an instructional questionnaire was administered to all students. Students taking the video-based course were asked to complete additional questions relating to the mediated learning experience.

Results of the Study

The results of the study can be summarized as follows:

1. There was no increase in the drop-out rate for students who took the video class. In fact the drop-out rate was far higher for students who took the class in the *conventional* classroom setting.

2. Percentage grades were higher for the students who took the course via videotape.

3. Students who took the video-based course, however, rated the course and the instructor significantly lower than those in the two comparison groups.

4. Students in both the video-based course and those who participated in the course during taping were significantly less satisfied with the course than those who participated in the course in the traditional classroom setting.

5. All students nevertheless rated the classes as being at least "good."

6. The graduate assistant in the video-based course was generally given high marks for her performance as the course facilitator.

7. A substantial number of students in the video-based course were dissatisfied with various aspects of the use of the medium.

Although the video delivered course appeared to be effective in terms of achievement and did offer students a viable, quality educational alternative to conventional instruction, there is little doubt that questions relating to how to optimally use the video medium remain to be resolved. Although student satisfaction was well within what we generally consider the "acceptable range," students in the video-based course expressed concerns about certain aspects of the course. Most of these concerns related to the use of the video medium.

Major concerns were:

1. It was difficult to see the monitor.
2. It was difficult to hear the audio.
3. Attention tended to wander.
4. There was less access to the instructor.

Feedback from the students provided the following explanations for the dissatisfaction that did exist:

1. Many students did not realize in advance the course was exclusively video-based.
2. Some students had not even seen in the course description that their section was video-based.
3. Students did not know what to expect from a video-based course.
4. Students did not know how to respond to the situation and how to get the most out of it.
5. Equipment was poorly selected and unreliable.
6. The graduate assistant did not know how to effectively cope with problems and equipment malfunctions.

Lessons Learned

Overall, the experiment was a successful one. We did show that video is a viable medium for the delivery of instruction at Boise State University and can effectively replace live faculty instruction. However, there are also many lessons that have been learned that will be useful for those contemplating using video-based instruction at our university and elsewhere.

The most important lessons we learned were the following:

1. **Technical excellence is important.** Students expect a high level of technical quality and are distracted when the production standards they have grown accustomed to in commercial television are not met. We have found that proper lighting, quality sound recording, and good shot selection are particularly important in educational video production. These production values will be given much more emphasis in future videotaping.

2. **Learners must be prepared for the video experience.** We had underestimated the difference between the videotaped course and the conventional course. We now realize the importance of providing an introductory session to introduce students to the video learning experience.

3. **We also learned how important variety is for maintaining learner attention.** Much more variety of learning activities will be engineered into future video-based courses.

4. **No matter how well produced the video and how interesting the subject matter, a video-based course is no better than the quality of the reception.** Therefore, in the future, we will put much more emphasis on what happens at the receiving site. The quality of the equipment, contingency planning, and graduate assistant training will figure much more prominently in our future efforts.

To assist others who wish to use video as an instructional medium, we have prepared a checklist, which should help others to use the video medium more effectively, by asking the right questions at the front-end of the project. This checklist is presented in Figure 1.

If we had had a checklist such as this one, we would have made many fewer mistakes and everyone (students, instructor, graduate assistant, and production staff) would have had a much more pleasant experience.

At Boise State University, we intend to use video even more widely as a cost-effective method of leveraging limited faculty resources. We have provided adequate evidence that video is a viable alternative to conventional instruction. However, we did find that the greatest barrier to effective video use in education (as in all innovations) is individual attitudes. There is little doubt that learning and retention are comparable (if not superior) through the use of video. We believe that student attitudes will become more positive toward the use of the medium, when the medium is utilized more effectively, that is, when:

1. Video-based courses are better planned up-front.
2. Students are more aware of their responsibilities in video-based learning.
3. Video quality is high.
4. More emphasis is put on successful reception and utilization at the receiving site.
5. Graduate assistants are properly trained to use the medium effectively.
6. The instructor maintains accessible office hours for students taking the video-based course.

Figure 1
Video Utilization Checklist

General questions:
1. Is videotape an appropriate alternative to live instruction?
2. How many learners need to be reached?
3. What is the level of the course?
4. What is the stability of the subject matter?
5. How often will the course be taught?
6. How much experience have the students had with video-based instruction in the past?
7. Are the students adequately prepared for the video learning experience?
8. What additional instructional resources are available?

Course design questions:
9. Is there enough variety in learning activities?
10. Have appropriate visuals been developed?
11. Is there enough redundnacy?
12. Are there adequate supplementary materials?
13. Are there provisions for ongoing monitoring and evaluation of the effectiveness of the course?
14. How accessible is the instructor?
15. Is there an adequate system for student access to the instructor?

Instructor questions:
16. Is the instructor comfortable as a video teacher?
17. Does the instructor have positive attitudes toward the use of the medium?
18. Does the instructor have a pleasing video personality?
19. Is the instructor well-organized?
20. Is the instructor willing to spend the time necessary for instructional planning?
21. How good is the instructor's eye contact with the TV camera?
22. How good is the instructor's voice quality?

23. How effective is the instructor's pacing/pausing?
24. How responsive is the instructor to the audience?
25. How well does the instructor regularly check up on student learning?

Technical quality questions:
26. How good is the quality of the video signal?
27. How good is the quality of the audio signal?
28. How effective is the lighting?
29. How appropriate are the visuals?
30. How good is the camera work?
31. How effective is the pacing of the production?
32. How well does the director cue the instructor?
33. How well does the production team work together?

Receiving site questions:
34. How clear is the television monitor image?
35. How clear is the sound quality?
36. Can students in all areas of the receiving site room see and hear clearly?
37. How dependable is the equipment?
38. Is there back-up equipment?
39. Is there a troubleshooting guide and a hot line to technical advice available at the site?
40. Is there a contingency plan?

Graduate assistant questions:
41. Is the graduate assistant familiar with the subject matter?
42. Is the graduate assistant familiar with the technology?
43. Does the graduate assistant have positive attitudes toward the medium?
44. Has the graduate assistant been adequately trained to use the medium?
45. Is the graduate assistant adaptable and cool under pressure?

As cost-effectiveness increasingly becomes an issue in higher education (Diamond, *et al.*, 1975), we strongly recommend the use of video to leverage instructional resources. However, as Spitzer (1986) has stressed in an earlier article in these pages, implementation has been (and still is) the missing link in educational technology. There is no substitute for planning, especially anticipating problems before they occur. By learning from **our** experiences, *you* will be able to do a much more effective job of utilizing video. □

References

Diamond, M. *et al. Instructional Development for Individualized Learning in Higher Education.* Englewood Cliffs, NJ: Educational Technology Publications, 1975.

Fleming, M. L., and Levie, H. W. *Instructional Message Design.* Englewood Cliffs, NJ: Educational Technology Publications, 1978.

Schramm, W. *Big Media, Little Media: Tools and Technologies for Instruction,* Beverly Hills, CA: Sage Publications, 1977.

Spitzer, D. R. Implementation: The Missing Link in Educational Technology. *Educational Technology,* September 1986, *26* (9), 47-49.

Some Advantages and Disadvantages of Narrow-Cast Television: One Instructor's Experience

David W. Dalton

Introduction

During my second semester at Indiana University, I was approached by my department chair and asked if I would like to narrow-cast the sessions of my introductory Instructional Development course to the "core campus" of IU. Here, the "core campus" can mean many things, depending on how and when the concept is applied. Generally, the core campuses of Indiana University include Indiana University Purdue University Indianapolis (IUPUI) and the Bloomington campus.

At Bloomington, the Instructional Systems Technology department has made a commitment to teaching various programs at the IUPUI campus, but due to factors including commute time and distance and the lack of qualified instructors in Indianapolis, we often have difficulty in meeting those commitments. In short, the Bloomington faculty cannot be in two places at one time.

Enter the Indiana Higher Education Telecommunication System or IHETS. IHETS is a narrow-cast instructional television system that allows us to be in several places at once through the "magic" of television. On the Bloomington campus, instructional television programming is accomplished through the Radio and Television department and cable-cast to the receiving classrooms at other campuses.

My first reaction to such a proposition was admittedly a bit egocentric. I felt as if I had suddenly been made a "star." So, despite some warnings from a colleague who had been televising such courses in the past, I agreed and plunged headlong into awaiting fame.

Some Disadvantages

First, the bad news. Needless to say, my naivety shone through almost immediately. My first task

David W. Dalton is Assistant Professor, Instructional Systems Technology, School of Education, Indiana University, Bloomington, Indiana.

was to tour the studio where the taping was to occur. To this day, I must say that I am still somewhat awed by the television technology. The studio used for this ITV course was shared by WTIU, the local public television station. Although not grand praise, these facilities are among the finest of their kind in the State.

The lights, cameras, and action abounding before me convinced me that I was indeed going to be a "star." My director, John Winninger, struggled hard to convince me that it really was not that big a deal, but nevertheless, I was awestruck. Together, we decided that we would narrow-cast live so that the audience at IUPUI could ask any questions as they arose. We also decided to include the entire Bloomington class as the studio audience and placed them on a lighted stage.

The first major disadvantage that we encountered seems, in hindsight, to be somewhat obvious. From the first night the studio lights and the TV apparatus seemed to stifle the questions of most of the students in the "live" audience. They seemed to feel as if their questions would make the production suffer. Despite many attempts, it was exceedingly difficult to sustain class discussions or encourage questioning. To rectify this situation, we moved the audience off-stage, and placed microphones overhead to pick up their questions. This change not only improved the amounts of discussion, but also improved my ability to engage in meaningful eye contact with the studio group. However, the frequency of questions still was far less than a "conventional" classroom environment and I noted an overall diminished sense of rapport with the studio group as the result.

On the subject of questions: because of budget constraints, the direct telecommunications link with the IUPUI audience was removed. Now, they had to pick up a device similar to a standard telephone which would actually place what amounted to a long-distance phone call for each question. Unfortunately, this calling mechanism required eight seconds to accomplish. Now, eight seconds may not seem like a great deal of time, but it always seemed to be a "just missed" situation. I'd call for questions and wait one or two seconds too few, or they would call and catch me as I began a new topic, one or two seconds too late.

Of course, there were minor inconveniences encountered all the time. For example, the need to dress and adopt a general demeanor that was more formal than my usual style was a bit of a nuisance at first, as was wearing the microphone and earpiece.

But the major problems surrounding this experience related to handling the logistics with the IUPUI group. Despite assurances to the contrary, there was never a reliable contact person in Indianapolis that would handle such seemingly trivial matters as proctoring and collecting exams, distributing class materials, collecting assignments, etc. For example, when it came time for the final exam, the students sat for two hours waiting for someone to show up with an explanation for why their study time had been so ill-spent.

One additional problem that caused difficulty throughout the semester was a calendar conflict. The official IUPUI semester began and ended 10 days before that of the Bloomington campus, and the spring recess for the two campuses fell during different weeks. The result was a condensing of the number of sessions and an occasional tape-delay for one or more of the audiences.

Some Advantages of ITV

Now for the good news. In many ways, this ITV experience proved beneficial personally and professionally. Although I've been teaching for many years, I've never been forced to prepare as carefully and thoughtfully as when the course was televised. Although my first reaction to the amount of planning required was that it would constrain my creativity and stifle spontaneity, not to mention detract from my other commitments, I found that the opposite was often the case. The more carefully I planned the presentations, the more comfortable I became in digressing and enlivening the material. By knowing what was to come next, I was able to relax and not worry about filling time or "flying by the seat of my pants."

From the ID perspective, this experience once again demonstrated the utility of carefully designed and developed instruction. I not only felt more comfortable with the content, but I also felt more effective as an instructor.

There were also advantages from a message design perspective. Instead of displaying transparencies or simply "talking through" important points, I was able to use computer-generated text to highlight important points, show the learners examples, display the lesson outline, etc. In addition, the visuals I developed were displayed with a close-up camera and a "chroma key" that allowed me to "zoom in" on each individual section of the graphic and drop back for a more holistic look. The manipulations of text and visual material were far more sophisticated than what I was ordinarily capable of accomplishing with a simple overhead projector or chalkboard.

Another notable advantage of this system was related to the taping itself. At the conclusion of each show, I received a standard VHS-format cassette which I could then check out to students who had either missed the class session or who needed additional review. As a footnote, in this particular class, approximately 50 percent of the learners are not native English language speakers and there is a significant range in their respective language abilities. These tapes provided many of these students with a review opportunity not usually possible. However, there were lingering delivery problems, principally due to the lack of public access viewing stations, that prevented this system from being used to its fullest.

Another advantage related to the video equipment was the capability of taping sessions ahead of time. On one occasion, I needed to attend a conference during the time when the class was held. Here, I simply arranged to tape the session a week earlier and narrow-cast it during the regularly scheduled time slot.

Of course, there were larger benefits than the technological miracles offered by the video equipment *per se*. First, the course experience was somewhat enriched through the addition of the IUPUI group. These learners were representative of a population of learners that are generally somewhat older and more focused in their motives than the Bloomington audience. Yet, the Bloomington crowd represents a very diverse and culturally heterogeneous mixture of learners. The two populations seem to complement each other well.

The benefits to the University seem somewhat obvious. Because an instructor can be several places at once, the costs, especially those incurred by the remote campuses, are low, limited almost entirely to very small administrative costs. In addition, the learners on the remote campuses benefit from courses that, because of small enrollments, high costs, or unavailability of faculty, would otherwise not be offered. Because of these factors, the attitudes of the IUPUI learners was extremely positive, despite many logistical problems.

Finally, the most salient advantage from my perspective was the opportunity to "practice what we preach" to a greater extent. As an instructional developer, it was far more credible for me to offer such a course through a more deliberate, mediated approach than the conventional classroom. This is not to say that this course was necessarily representative of the optimal ID environment; certainly, it was not. However, in a larger sense, the learners in this course were exposed to some media and design options that they ordinarily would not see.

The Balance Sheet

Would I do it again? Probably. However, the most frustrating problem and the reason why I may be much less eager to undertake the instructional television experience again in the future is the attitude displayed by the local group of students. With few exceptions, these learners responded with the attitude that this technology, no matter how beneficial it may be to learning here and throughout the system, somehow cheated them out of the best possible learning environment because of its "staged" nature. Instead of being willing to experiment with a new type of learning situation, many seemed to feel compelled to demand the security and comfort of the conventional classroom.

Among all the lessons that I learned from this experience, one far out-shone the others in my own understanding of the field of instructional development: we as professionals must advocate strongly for pedagogical change, or instructional interventions, no matter how ultimately beneficial we may believe they are, will continue to be viewed merely as new "teaching toys"—here for today, but gone tomorrow. In short, before we can indeed practice what we preach, we must actively sell the better mousetrap. □

An Apple a Day and at Night: A Distance Tutoring Program for At-Risk Students

Steven M. Ross, Lana Smith, Gary Morrison, and Ann Erickson

A critical role for computers in education is helping to reduce the number of "at risk" children in our schools. Currently such children are estimated to comprise as much as 35 to 40 percent of the K-12 population in the United States ("Technology and the At-Risk Student," 1988). Many different types of individuals fall into the at-risk category, but common profiles include some combination of being black or Hispanic, economically disadvantaged, low achieving, and physically or mentally handicapped. Importantly, regardless of background, all share the characteristic of being significantly less likely to complete school compared to other students.

On the surface and seemingly in practice, computer-based instruction (CBI) offers the at-risk student several important types of benefits ("Technology and the At-Risk Student," 1988). One is the development of increased self-efficacy or self-esteem through actively controlling the computer to support one's own learning. A second is the individualization received through the branching and self-pacing features of CBI programs. A third is the opportunity to work in a nonjudgmental context where successes and failures are private (i.e., between the student and the computer). A fourth is receiving direct instruction and remediation in basic skills to supplement regular classroom learning. For the most part, all of these benefits seem realizable through students' successful interactions with well-designed CBI and applications software. Taking a somewhat different orientation is a fifth type of intervention in which furthering opportunities to interact with *human* teachers is the major goal. This orientation, called "distance tutoring,"

Steven M. Ross, Lana Smith, and Gary Morrison are faculty members at Memphis State University. Ann Erickson is with the Memphis City Schools, Memphis, Tennessee.

is the focus of the special program for at-risk minority students to be described in this article.

Program Context and Objectives

The context for the distance tutoring program is the Memphis Apple Classroom of Tomorrow (ACOT) project, a collaborative effort between Apple Computer, Inc., the Memphis City Schools, and Memphis State University. In the fall of the 1986-1987 school year, the target elementary school in Memphis received a grant from Apple Computer to equip fifth- and sixth-grade classrooms with microcomputers. A key feature that distinguishes this school from the 12 other ACOT sites nationwide is its nearly 100 percent minority student population and location in a poor inner-city neighborhood. A standard component of the ACOT orientation is to provide each child with one computer to use at school and another to use at home, thus allowing for virtually unlimited computer access to support instructional (CBI) and tool applications (e.g., word-processing). Based on the inferred special needs of its at-risk student participants, unique to Memphis ACOT is the assignment, to each child, of a personal tutor who is a Master of Arts in Teaching (MAT) candidate at Memphis State University. The tutor leaves assignments and writes messages over an electronic Bulletin Board System (BBS) accessed by modem.

The rationale for the BBS program is based, in part, on recognition of personalized tutoring as a highly powerful strategy for adapting instruction to individuals (Bloom, 1984; Slavin, 1988). Unfortunately, the realities of limited school budgets and large class sizes make such tutoring highly impractical in all but a few situations (Eisenberg, Fresko, and Carmeli, 1983). Even where qualified adults may be willing to tutor children on a voluntary basis, the main barrier is the considerable time, expense, and logistical difficulties involved in making face-to-face contacts. Although a distance tutoring program cannot be expected to convey the full impact of conventional tutoring, it can seemingly offer many of its advantages without its practical constraints.

Type of Distance Learning Systems

Current applications of distance learning use two basic forms of telecommunications, text teleconferencing and electronic mail (Quinn, Mehan, Levin, and Black, 1983). The main distinction is that teleconferencing, but not electronic mail, takes place in "real time," as a telephone conversation does. Participants "speak" by typing messages on a computer terminal which are transferred by modem to the receiver's computer; "listening" takes the form of reading the messages as they appear on the monitor. In contrast, electronic mail systems operate in the same way as the exchange of written letters or memos. Once the electronic messages are transmitted to another person, they are stored in a BBS file ("mailbox") under his/her name, to be read and responded to at the recipient's convenience.

Each of these orientations offers different types of advantages for distance tutoring. Teleconferencing permits the tutor to interact directly with the student by asking questions, presenting explanations, giving feedback, and so on. It also provides a way in which students attending schools in different geographical areas can regularly converse with one another. Sayers and Brown (1987), for example, describe a bilingual project between "sister classes" in the United States, Mexico, and Puerto Rico. By communicating with students from foreign countries, students felt they were learning a living language rather than memorizing grammar. Butler and Jobes (1987) cite similar goals in a link between Australian and American schools emphasizing the exchange of cultural information (also see Benson and Hirschen, 1987).

In contrast to real-time exchanges, electronic mail creates a relatively permanent written communication that students can access at their convenience, read and reread at a comfortable pace, and respond to at a future time. Quinn *et al.* (1983) used such a system in teaching a college-level course on instructional strategies. They concluded that a non-real time message system evokes a very different pattern of questioning and answering (e.g., longer, more detailed responses) than do face-to-face interactions, and thus can serve as a valuable supplementary means of teaching a course. Despite the positive experiences that have been reported in using such systems (Davis, 1987; Zellers, 1987), there are also disadvantages (Newman, 1987). One is obviously the delay in communications between the teacher and the student. Another is that students must be sufficiently disciplined and knowledgeable to access the messages in a timely manner, interpret and complete the assignments on their own (or seek help), and transmit their responses to meet deadlines. A third is that the lack of physical interaction creates a learning situation that is relatively passive and isolated (Jones and O'Shea, 1982).

For maximum flexibility in distance tutoring, it would normally be desirable to incorporate features of both electronic mail and teleconferencing. Unfortunately, only electronic mail was an option in the Memphis ACOT project due to the capabilities of the available BBS hardware and software. As will be described, it was therefore necessary to design tutoring methods and assignments to fit within the constraints of that environment. In addition to the BBS component, a key program emphasis was the development of a personal relationship between tutors and tutees (McConnell, 1986). In many distance learning programs, including school telephone tutoring projects or "homework hotlines" (Pedley, 1987; Wood, 1986), the tutors and tutees interact essentially as strangers.

The Human Factor: The ACOT Tutors

Even with the most technologically advanced communication devices, the success of a distance tutoring program will depend ultimately on the commitment and talents of its human tutors. For the initial year of the ACOT program, 1986-1987, high school students who were both high achievers and computer buffs were recruited. For their services, they received a computer and modem to use at home, and the opportunity to work with the new software and technology made available through the ACOT project activities. The high school students brought youthful enthusiasm and impressive computer expertise to their work, but a limitation that rapidly became apparent was their lack of training in teaching and motivating elementary school children. As teenagers, many also lacked the maturity and experience for managing time effectively and keeping the tutees on task. Due to this factor and the lack of time to prepare appropriate tutoring assignments, use of the tutoring component was consequently limited during this first year. Important products of the tryout, however, were the development of a working delivery system for the tutoring model as well as increased confidence that the model would be well-received by and helpful to students.

For the following year (1987-1988), it was decided that college students, being older and more experienced, would be better qualified to serve as tutors. Consistent with this idea, college students are currently being recruited as part of a recently formed national higher education coalition to help at-risk youths (Hirschorn, 1988). MAT candidates at Memphis State University were a natural choice for the present program given their graduate status and direct interests and training in teaching. Working as tutors was also expected to offer them a number of attractive benefits. One was to learn from firsthand experiences about computers and their uses in schools. Another was to broaden their professional development by working with the at-risk minority children in ACOT. A third was to obtain a home computer that they could use for personal work during the school year. A fourth was the opportunity to conduct their Master's thesis research on one of the numerous ACOT-related topics. With these built-in incentives, there were more volunteers than could be accommodated. The final group of 10 tutors consisted of four males (one minority) and six females, ranging in age from 23 to 47. Each served as a tutor for 2-3 students over the school year.

The BBS System

As previously indicated, the distance tutoring model operated via the exchange of electronic mail. The host system for the BBS was a 512K Macintosh with a 20 MB hard disk drive and an Apple Personal Modem. One dedicated phone line was used for all incoming calls. Students and tutors accessed the BBS using an Apple IIc computer, an Apple Personal Modem, and Apple Access II software. Once on-line, they entered their personal password and were then acknowledged by the system. Next, any messages that had been sent to them since their last access were listed by sender's name and message number. They could then select between menu options allowing them to read, post, or delete messages. Students primarily communicated over a "public" bulletin board, but they could also access a restricted section to send private messages. Tutors had access to both of these boards as well as to a third, "Modem Tutor," section for posting messages that could be viewed only by other tutors and ACOT staff. In posting a message, the user entered the recipient's name and a brief message title. After typing the message, he/she could choose to send it as it appeared, edit it, or delete it. File transfer was not easily accomplished and therefore was not used to compose upload messages created offline. Generally, the system functioned smoothly throughout the year, with the exception of occasional down time during electrical storms and the breakdown of a few modems.

Tutoring Methods and Assignments

The initial phase of the tutoring program involved conducting a series of orientation and training workshops for students, parents, and tutors. These sessions gave basic demonstrations of how to connect and operate the computer equipment. A second part of the training dealt with procedures for operating the BBS. Because this BBS was completely menu-driven and very user-friendly, its procedures were easy to follow.

Considering that electronic mail would be the primary tutoring mode, it was decided to orient assignments around writing skills activities that would depend on the exchange of communica-

tions over time rather than through real-time interactions. The initial assignment asked students to construct a "contrast paragraph" comparing their own background and interests to those of their tutors. One opportunity to collect the personal data was provided during an early face-to-face social gathering with the tutors. Later on, students could request any needed information in messages transmitted over the BBS. Completed paragraphs were posted on the BBS for the tutors to examine and react to through comments and feedback. Later, a printed copy of the assignment was handed in and evaluated by the classroom teacher. Additional BBS assignments during the first semester were patterned on this basic orientation.

Assignments during the second semester were used to help students expand their reading and writing vocabulary. Tutors were given copies of vocabulary journals which their tutees had created using an AppleWorks data base. The journals contained four categories of information: new words, context sentence, definition, and original sentence. Exercises created by tutors listed five of the words and asked the student to use each correctly in a sentence, identify synonyms, and indicate correct or incorrect uses in completed sentences. Tutors evaluated this work and submitted the grades to the classroom teacher. In addition to these and related formal assignments, tutors and tutees were strongly encouraged to use the BBS informally to share experiences and maintain regular contact.

Research Outcomes

A key element of the ACOT project is ongoing research to identify activities and outcomes associated with its component programs. With regard to achievement, results on standardized tests have shown some advantages for ACOT students relative to matched control groups (Kitabchi, 1987, 1988). These results need to be viewed cautiously, however, due to the many confounding variables, including Hawthorne effects, that could have contributed to them. More informative findings for understanding ACOT's impact have come from localized examinations of individual programs. Three Master's thesis studies, supervised by the present authors, were specifically directed to the distance tutoring program. Areas of interest were BBS activities, tutor roles, and writing skills development.

BBS activities. To determine how the BBS was used by participants, Ulrich (1988) tabulated and coded every message sent over three one-week periods dispersed across the school year. She found that overall use was extensive, with the number of messages totaling 231, 131, and 205 during the respective periods. Active users consisted of 10 tutors, 5 ACOT teachers and staff, and 16 students. Each of these three groups accounted for approximately equal proportions of the total messages posted and received, although the activity rate per person tended to be lower for students than for the adults. An interesting finding was the clear tendency for girls to use the BBS more frequently than boys ($p < .001$). Of all student messages sent, 87 percent were by girls whereas only 13 percent were by boys. Also, while boys sent no messages to other students, girls sent 21 percent ($n = 33$) of their total to peers, of which all but two messages were to other girls. Thematic classification of messages showed that social messages accounted for 48% of the total, and academic messages, the next most frequent type, accounted for 15%. Other, lesser used categories were tutor activities, thesis activities, announcements, exercises/explanations, grade reports, and technical assistance requests or replies.

In reacting to their experiences, students were positive about the tutoring program in general, but discouraged about the operational limitations of the BBS which made access difficult and communications with tutors less regular and frequent than desired. Another discouraging (but expected) outcome was that one-third of the students exhibited little BBS usage. In several cases, equipment breakdowns were the problem. In other cases, the main reasons were inability to access the system at convenient times, limited opportunity to use the telephone at home, or simply lack of motivation to complete assignments. Overall completion rates for BBS assignments remained higher throughout the year than those for regular homework, but still the unrealized (and perhaps unrealistic goal) had been to involve *every* child as an active participant.

Tutor roles. A comprehensive examination of the tutoring experience was made by Parry (1988) using data from tutors' daily logs, an end-of-year attitude survey, tutor interviews, and observations of his own tutoring activities. Findings indicated that, overall, tutors regarded their activities as rewarding due to helping at-risk children to learn and being part of an innovative program. In particular, they felt that their positions as role models and mentors were important in setting positive examples for the children and in motivating them to achieve. The fact that the written word was the sole means of communicating was perceived to be a strong advantage in helping to improve reading and writing skills.

Similar to the students, tutors also expressed dissatisfactions with the BBS operations. The sin-

gle line connection to the BBS was frequently busy, forcing them to call many times to gain access. The inability to communicate on-line precluded real-time interchange of messages and feedback. Restrictions in the time (30 min.) and space allotted for writing messages often made communications pressured and unnatural. Separate from these procedural factors, tutors noted the absence of a systematic curriculum plan for integrating tutoring and classroom learning. Some assignments (writing/correcting essays) appeared to work well on the BBS, while others (vocabulary exercises) might have been more efficiently handled through conventional means. Despite these limitations, tutors saw great promise for the program and indicated that they would enthusiastically volunteer again.

Writing skills. Recognizing that ACOT students were receiving extensive practice in writing, both formally in BBS writing assignments and informally in composing BBS messages, Woodson (1988) questioned whether writing skills might improve as a result. In her study, the ACOT group was compared on a variety of writing criteria to a control class attending the same grade at the same school. Writing samples from alternative assignments (essay and letter) were collected from both groups at two times during the year. Analyses were made by independent judges of 19 performance variables, including number of words, topic sentence use, spelling, subject/verb agreement, redundancy, tense shifts, run-ons, and fragments. Results indicated that the ACOT students performed better grammatically and structurally across the two writing samples. They wrote more; used more topic sentences; had better coherence and unity; and made fewer errors in tense, spelling, and capitalization. The control group's writing samples, however, reflected more details and better usage of pronouns. In interviews, teachers and tutors concurred that the BBS activities were highly beneficial for teaching writing, but expressed the same dissatisfactions noted in the aforegoing studies regarding the limitations of the BBS operations and interface.

Conclusions

Uses of the distance tutoring program during the past two years have demonstrated the potential of this type of intervention to help at-risk children learn and gain self-confidence. An important accomplishment was using technology to bridge the physical and cultural distance that existed between the two diverse groups of participants, the middle-class college-educated tutors and the minority tutees. The idea that college students would make willing and competent tutors (Hirschorn, 1988) was also clearly supported (see Parry, 1988). Aside from its academic applications, the BBS proved highly popular as a vehicle for informal (social) communications among and between students, teachers, and tutors (Ulrich, 1988). Its stimulation of natural uses of writing for these purposes appears to have contributed to the superior performances of the ACOT group in Woodson's (1988) study.

We also learned that, as with other types of computer-based learning (CBI, CMI, or interactive video), new instructional technologies cannot simply be interjected into a classroom but must be systematically integrated with the existing curriculum and teaching methods. To make distance tutoring work well with at-risk children, we propose that attention be given to the following conditions:

1. Distance learning must be personalized and humanized (McConnell, 1986; Paulet, 1987). Accordingly, frequent occasions for tutors and tutees to meet face-to-face and become better acquainted are needed throughout the year.

2. Tutors should not only be knowledgeable about content, but also be able to teach or explain it to others. Experience in working with children is another highly desirable attribute. Assuming the likely inavailability of professional teachers, college education majors seem ideal for tutoring roles.

3. The BBS operation and interface must facilitate effective tutor-student interactions, not make them unnatural or difficult.

4. Tutoring assignments must be carefully matched to regular classroom learning activities and to the delivery capabilities of the BBS.

5. Since the locus of distance tutoring is the home, more extensive and systematic involvement of parents is needed.

6. Tutoring roles should be consistent and not intermittently alternated between that of mentor (helping, supporting) and evaluator (assigning points and grades).

With these considerations in mind, this year's program (to start in January, 1989) will employ the new Apple-Link Personal Edition to create a more powerful and flexible BBS. A major advantage will be multiple user access via Telenet or Tymnet dial-up networks. Others will be capabilities for real-time interactive responding, uploading/downloading of messages, "chat rooms" for group tutoring sessions, and access to numerous Apple-Link information sources (encyclopedia, news, etc.). Tutoring assignments have been newly designed to achieve a better fit with the curriculum and BBS capabilities. Increased efforts to involve parents and to arrange more face-to-face contacts between students and tutors are expected to build

a stronger sense of community among participants. With the continued evaluation and refinement of the program, our understanding has increased of how to integrate effective instructional design and delivery methods within a distance learning environment. The important payoff is seeing children who are presently "at risk" become less likely to remain that way through the intervention of an innovative and powerful technology. □

References

Benson, G. M., and Hirschen, W. Long Distance Learning: New Windows for Technology. *Principal*, 1987, *67*, 18-20.

Bloom, B. S. The 2 Sigma Problem: The Search for Methods of Group Instruction as Effective as One-to-One Tutoring. *Educational Researcher*, 1984, *13*, 14-16.

Butler, F., and Jobe, H.M. The Australian-American Connection. *The Computing Teacher*, 1987, *14*, 25-26.

Davis, S. Teleteaching Project Links Remote Iowa Districts. *Electronic Learning*, 1987, *6*, 16.

Eisenberg, T., Fresko, B., and Carmeli, M. A Follow-Up Study of Disadvantaged Children Two Years After Being Tutored. *Journal of Educational Research*, 1983, *76*, 302-306.

Hirschorn, M. W. Coalition of 120 Colleges Hopes to Encourage a Million Students to Tutor "At-Risk" Youths. *Chronicle of Higher Education*, January 1988 pp. 35, 38.

Jones, A., and O'Shea, T. Barriers to the Use of Computer Assisted Learning. *British Journal of Educational Technology*, 1982, *13*, 207-217.

Kitabchi, G. *Final Report for the Evaluation of the Apple Classroom of Tomorrow: Project Phase II.* Memphis, TN: Memphis City Schools Division of Research Services. June, 1987.

Kitabchi, G. *Final Report for the Evaluation of the Apple Classroom of Tomorrow: Project Phase III.* Memphis, TN: Memphis City Schools Division of Research Services. June 1988.

McConnell, D. The Impact of the Cyclops Shared-Screen Teleconferencing in Distance Tutoring. *British Journal of Educational Technology*, 1986, *17*, 41-74.

Newman, D. Local and Long Distance Computer Networking for Science Classrooms. *Educational Technology*, 1987, *27* (6), p. 20.

Parry, M. S. *A Descriptive Analysis of the Attitudes of Tutors Involved in the Memphis ACOT Modem Tutoring Program.* Unpublished Master's Thesis, Memphis State University, 1988.

Paulet, R. Counseling Distance Learners. *Tech Trends*, September 1987, *32*, 26-28.

Pedley, F. Homework Hotline. *Times Educational Supplement*, 3693: April 24, 1987, *19*, p. 19.

Quinn, C., Mehan, H., Levin, J. A., and Block, S. D. Real Education in Non-Real Time: The Use of Electronic Message Systems for Instruction. *Instructional Science*, 1983, *11*, 313-327.

Sayers, D., and Brown, K. Bilingual Education and Telecommunications: A Perfect Fit. *The Computing Teacher*, 1987, *14*, 23-24.

Slavin, R. E. *Educational Psychology: Theory Into Practice.* Englewood Cliffs, NJ: Prentice-Hall, Inc. 1988.

Technology and the At-Risk Student. *Electronic Learning,* November/December 1988, *8,* pp. 36-39, 42-49.

Ulrich, L. M. *A Study of Electronic Bulletin Board Usage by Participants in the Memphis ACOT Program.* Unpublished Master's Thesis, Memphis State University, 1988.

Wood, J. School Based Homework Assistance. *Educational Horizons,* 1986, *64,* 99-101.

Woodson, M.E. *An Evaluation of the Writing Skills of Sixth-Grade Students Participating in the Apple Classroom of Tomorrow Project.* Unpublished Master's Thesis, Memphis State University, 1988.

Zellers, B. Telecommunications, Apple, and Hobart City Schools. *The Computing Teacher,* 1987, *14,* 33.

Telecommunications Skills Training for Students with Learning Disabilites: An Exploratory Study

Barbara J. Edwards and Mark A. Koorland

One promising advance in computer technology applications has been the development and use of telecommunications. It is believed that telecommunications can provide a medium for countless instructional applications by providing students with opportunities to communicate with other people via computers (Adams and Bott, 1984). While microcomputers have been found to be useful instructional tools in special education classrooms (Gall, 1985), the benefits of newer telecommunications technology for instructional purposes have not been widely explored.

Electronic telecommunications typically involve transferring information from computer to computer via telephone lines. This linkage is possible because both computers and telephones use electricity to communicate (Archibald, 1983). Four components are necessary for electronic telecommunications: (a) computers, (b) modems, (c) telephone lines, and (d) telecommunications software.

With acquisition of compatible hardware and software, it is possible to connect directly with another computer or with an electronic network. Networks offer information of interest to users in either of two formats: electronic bulletin boards or electronic databases (Raimondi, 1984).

The electronic bulletin board has been likened to its non-electronic counterpart (Campbell, Gibbs and Snodgrass, 1982). A user can "post" information electronically on topics, and other users can access that information. Messages posted on electronic bulletin boards are referred to as electronic mail.

Electronic databases constitute one of the largest sources of information available today (Bakke, 1984). Of the various kinds of databases, the general information type, containing a variety of general information, is typically easiest to use (Neumann, 1982).

Barbara J. Edwards is a doctoral candidate in Special Education at the University of Kentucky in Lexington. **Mark A. Koorland** is Professor, Special Education Department, Florida State University, Tallahassee.

Students have been taught to use telecommunications to access information on a variety of topics for research purposes. Information sources such as the new Prodigy service, for example, provide current information about news, weather, sports, etc. One telecommunication service even offers an on-line encyclopedia for information access. Bulletin boards with electronic mail can provide students with opportunities to write about their interests in a new and exciting format.

Mildly handicapped students typically are required to perform increasingly complex language arts and writing activities in secondary programs. Yet, mildly handicapped learners, especially those with learning disabilities, besides demonstrating deficient language strategies (Sitko and Gillespie, 1978), are often characterized as inactive learners (Torgesen and Licht, 1984). Consequently, students with learning disabilities are frequently provided academic tasks recognizably less difficult than required of non-handicapped peers. Fewer skills and an awareness of being different often make LD students unwilling to participate in academic tasks of many kinds. Additionally, these students encounter failure to such a marked degree during their schooling that they often require application of involved motivation strategies in order to perform in school. Clearly, learning activities with some degree of sophistication and motivational properties are worthwhile for mildly handicapped students. For example, writing is a valued communication ability. One way to improve one's own writing is to interact with responding readers (Reid, 1988). Exchange of written material using telecommunications could assist in motivating students to write.

There is quite limited information about teaching secondary students with mild handicaps the relatively sophisticated skills associated with telecommunications, and no studies indicating the practicality or effectiveness of such training. Nevertheless, telecommunications skills offer a motivating medium for a wide range of applications. The objectives of the present study were to determine if basic tasks for telecommunications could be learned without great difficulty by secondary students with learning disabilities, and if students would react favorably to telecommunications instruction.

Method

Subjects and Setting. Five students in a self-contained class for students with learning and behavior disorders in an urban senior high school participated. All students had previously used a computer for playing drill and practice games or for word processing. One student had created graphics. None had programming or telecommunications experience. Typing skill, across the group, ranged from good to letter-by-letter typing. Ages ranged from 15 to 18. Reading level grade equivalent scores ranged from 3.4 to 6.8 with a mean of 4.56.

Instruction took place in a classroom at a table near a phone connection. Students sat at chairs close to the table in order to see the monitor and other equipment. Regular classroom activities were conducted in the opposite part of the room.

Equipment. The hardware used included: (a) one Apple IIe microcomputer with 64K RAM, (b) two Apple UniDisk drives, (c) one Apple monochrome monitor, and (d) one Apple 300/1200 baud Personal Modem. The terminal software program was Apple Access II (Apple Computer, Inc., 1983). A local bulletin board system for use by local special education teachers and administrators was available. Students were restricted to one of eight bulletin boards on the system. Use of a local board eliminated long distance fees and permitted the investigator to monitor log-on times and to read students' messages.

Instrumentation. A task analysis of skills necessary for three basic telecommunications tasks, logging-on, sending, and receiving messages, was developed into a recording checklist. The checklist permitted verification, for each student, of either independent skill mastery or performance that required assistance. This checklist was submitted to a panel of six special education microcomputer specialists at the University of Kentucky to verify content validity. Additionally, interview questions for use at the end of training were developed to assess students' reactions to learning telecommunications.

Procedure. Students received eight, 55-minute class periods of instruction and practice over eight days. Students received and sent electronic mail to each other, their classroom teachers, and a school administrator. The three basic telecommunications tasks and associated skills taught to students are listed in Table 1.

The skills shown in Table 1 were taught by first describing hardware and software, discussing proper computer and disk handling, and describing the local bulletin board. Every instructional period thereafter was begun with review and demonstration of the skills taught the day before. Lessons ended with students performing the skill individually. Logging-on was taught the second instructional day. Sending messages was taught on Day 3, and receiving messages was taught on Day 5. During the remaining days, students practiced all skills individually. Students were evaluated for mastery on the individual steps of telecommunications on the last day of instruction.

Table 1

Telecommunications Skills

Logging-On Procedures
Insert disks in appropriate disk drives.
Turn on the computer.
Use the auto dial feature of the software program to access the bulletin board.
Use correct log-on procedure.
Enter a password.

Sending Electronic Mail
Access correct bulletin board.
Access enter message feature.
Enter the subject of the message.
Enter name of the receiver of the message.
Enter the message.
Save the message.
Quit the system.

Receiving Electronic Mail
Log-on.
Retrieve messages.
Quit the system.

Results

A percent score was determined for each student, based on the number of steps independently mastered within each basic telecommunications task. Scores for logging-on ($\bar{x}$=100 percent mastered), sending ($\bar{x}$=90 percent mastered), and receiving ($\bar{x}$=100 percent mastered) electronic mail, were compared using the Cochran Q test for related samples (Siegel, 1956). The comparison was used to determine if students as a group encountered more difficulty in performing any of the three telecommunications tasks over another. There were no significant differences between scores on the three tasks (Q=4.00, df=2, p=.15). Two students, however, required assistance on a few steps involved in entering a message during the message sending task.

At the study's conclusion, students were queried about their reactions to telecommunications. All students agreed that they would like their friends to use telecommunications. Four of five agreed that they enjoyed using telecommunications, they felt comfortable using it, and they wished they could use telecommunications to do more assignments. All five disagreed that they had too much time at the computer, and four disagreed that using telecommunications was boring. Only two, however, agreed that learning telecommunications was easy.

Discussion

Results suggest that learning-disabled students in this study could learn basic skills necessary to take advantage of telecommunications technology. Students generally liked telecommunications, although they indicated it was difficult to master. The need for occasional teacher assistance noted after students were trained could probably be reduced by more frequent telecommunications use.

There are a number of limitations to this study. First, the study was exploratory and would require replication to determine the durability of the particular instruction and task analysis employed. Second, instruction took place on one computer. Consequently, students waited turns to practice and this could have resulted in less independent performance while sending messages, noted in two students. Researchers might explore the effectiveness of different instructional strategies for telecommunications. Also, it is not known whether younger students can or should be taught to send and receive electronic mail. Further study is needed to determine the most beneficial instructional applications of telecommunications for students with mild handicaps. The motivating features of telecommunications noted in student responses to this study may assist learning disabled students with written communication, and help orient them to technology. □

References

Adams, D.M., and Bott, D.A. Tapping into the World: Computer Telecommunications Networks and Schools. *Computers in the Schools*, 1984, *1* (3), 3-17.

Apple Computer, Inc. Apple Access II (*Computer program*). Cupertino, CA: Apple Computer, Inc., 1983.

Archibald, D. What Is and What's to Come in Telecommunications. *Softalk*, January 1983, 185-189.

Bakke, T.W. *Existing and Emerging Technologies in Education: A Descriptive Overview* (CREATE Monograph Series). Palo Alto, CA: American Institutes for Research in the Behavioral Sciences, 1984.

Campbell, B., Gibbs, L., and Snodgrass, G. Telecommunications: Information Systems for Special Education's Future. *Journal of Special Education Technology*, 1982, *5* (1), 1-11.

Gail, R.S. *Keys to the World: A Brief Examination of Microcomputer Technology for Telecommunications*. San Francisco, CA: Annual Convention of the Association for Children and Adults with Learning Disabilities. (ERIC Document Reproduction Service No. ED 258 395), 1985.

Neumann, R. Data Banks: Opening the Door to a World of Information. *Electronic Learning*, February 1982, 56-60.

Raimondi, S.L. Electronic Communication Networks. *Educational Technology*, 1984, *24*, 39-40.

Reid, D.K. *Teaching the Learning Disabled: A Cognitive Developmental Approach*. Boston: Allyn and Bacon, 1988.

Siegel, S. *Non Parametric Statistics for the Behavioral Sciences*. New York: McGraw-Hill, 1956.

Sitko, M.C., and Gillespie, P.H. Language and Speech Difficulties. In L. Mann, L. Goodman, and J.L. Wiederholt (Eds.), *Teaching the Learning Disabled Adolescent*. Boston: Houghton Mifflin, 1978.

Torgesen, J.K. and Licht, B.G. The Learning Disabled Child as an Inactive Learner: Retrospect and Prospects. In J.D. McKinney and L. Feagans (Eds.), *Current Topics in Learning Disabilities. Vol. 1.* Norwood, NJ: Ablex, 1984.

U.S. Congress, Office of Technology Assessment. *Linking for Learning: A New Course for Education*, OTA-SET-430. Washington, DC: U.S. Government Printing Office, November 1989.

Live from Germany: A Foreign Language Encounter Via Satellite

Hildburg Herbst and Peter Wiesner

On one day last fall, twenty undergraduate German language students at Rutgers and ten at Dickinson College "met" officials and students at Heidelberg University via satellite. For one hour, they took part in an experiment with telecommunications in foreign language instruction that permitted them to share in the excitement of the 600th anniversary celebrations at Heidelberg University.

The price tag for the experiment, all donations and free services included, was estimated at roughly $30,000, no doubt exceeding the cost of sending the same American students on an all-expense paid trip to Heidelberg during which cultural exposure and language skills were likely to have been nurtured through first-hand experiences.

Was it worth it? Our colleagues in the Rutgers Department of Germanic Languages posed that question to Dr. Thomas Naff, director of Intel-Ed, the organization which obtained the funding and made the arrangements to hook up Rutgers and Dickinson students with native speakers in Germany. According to Dr. Naff, the purpose of the project was to use telecommunications as a way of providing international exposure to the homeland bound. We were treated to a vision of technological progress in which national boundaries dissolved in video. Tomorrow, no sooner said than done, would bring a day when trans-Atlantic video would be as cheap as a telephone call. Nothing to lose, we suspended our disbelief and decided to give the experiment a try.

Manifest Technology

West German television handled the arrangements and covered the costs for originating the video in Germany and feeding video signals which were routed via Bundespost land lines to Intelsat and then down-linked in the United States via Comsat. The transmission cost of the video, covered by the Intel-Ed grant, was valued at $10,000 per hour, which included Bundespost land lines, uplinking, and transponder time on Intelsat and Comsat. The remaining costs, estimated at $20,000, included organizing the event, travel to Germany, video production, and downlinking.

As a one-way video conference with audio return, the students in the United States could see and hear the video from Germany. However, the participants in Germany could only hear the American students by phone. Photos of the American students were mailed before the transmission so that the Germans could "see" whom they addressed. The American students were also to see the photos of themselves inserted in the video from Germany. In the actual transmission, this proved to be more difficult than anticipated, and some heated discussion concerning these matters ensued on the coordinating telephone line between the American instructor and the representative from Intel-Ed who went to Heidelberg to arrange the details of the transmission.

Satellite Ahoy

The transmission with Heidelberg University was the third of a series of one-hour satellite sessions designed to expose American students to German language and culture.

The first transmission, involving only Dickinson College*, simulated a classroom in which the curatorial staff at the Bremerhaven Maritime Museum gave a lesson to the American students about the history of the German emigration to the United States. A video segment on the Maritime Museum was followed by the Germans quizzing the American students on assigned readings concerning the German emigration.

Since many of the Dickinson students were beginning speakers of German, they struggled to comprehend and express themselves. The Germans seemed to respond to the American "language gap" with impatience. Apparently they found it hard to believe that American college students were so far behind their Europeans counterparts in the mastery of a foreign language. It didn't help matters when the students heard themselves twice due to the failure of controlling audio feedback at the Bremerhaven location.

Hildburg Herbst is Associate Professor, Department of Germanic Languages and Literature, Rutgers University, New Brunswick, New Jersey. **Peter Wiesner** is Coordinator of Academic Television, Office of Television and Radio, Rutgers University.

*For a thorough account of the first transmission, see: Beverly D. Eddy, Live from Germany: A Look at Satellite Instruction. *Die Unterrichtspraxis*, Fall 1986, *19*(2), 213-219.

The second transmission of one full hour included ten Rutgers University undergraduates. The first and third twenty-minute segment was reserved for Dickinson, and the second segment for Rutgers. This transmission was more successful than the first one in terms of technology, pedagogy, and diplomacy. More advanced students were recruited for the venture. Audio feedback was eliminated by headsets at the German location. Also, this time the stiff didactic exchange of a typical classroom lesson was avoided by inviting the German rock star, Klaus Lage, for the exchange in German. Because of the high-interest subject area, students were able to muster the courage and skills to sustain a dialogue.

In the third one-hour transmission, the first and third twenty-minute segments were reserved for Rutgers, and the second for Dickinson. In planning this transmission, we were faced with the Hegelian challenge of effecting a synthesis of a didactic exchange and natural dialogue. Intel-Ed invited the Rutgers and Dickinson colleagues to discuss how students might gain the most from the third experiment.

We educators wanted the exchange to be natural and informal. We didn't want our students to be buffaloed by the Heidelberg officials obviously intent on promoting Heidelberg's 600th anniversary celebration. We wanted them to be full-fledged participants in this telecommunications extravaganza. We wanted the media event to be a non-media event as well.

We reminded the staff at Intel-Ed to caution Herrn Doktor zu Putlitz (President of Heidelberg University), Herrn Doktor Must (director of public information), and Herrn Doktor Nordheim (a language professor and director of public information at Heidelberg) to avoid addressing our students from Mt. Olympus. We wanted to make sure that enough time would be allocated to allow our students to exchange views with *students* at Heidelberg.

Promotion Precedes Essence

After an hour of technical preparations to make sure that we were sending and receiving clear video and audio, the video feed from Heidelberg showed the sumptuous baroque interior of the "Aula" at Heidelberg University. Here was old world splendor in contrast to the stark surrounding in which the American students at Rutgers found themselves—an old army barrack, formerly part of Camp Kilmer, housing a television studio set that had been jazzed up in *Miami Vice* colors by the Rutgers

Office of Television and Radio for another program.

Rektor zu Putlitz lectured for about 15 minutes about the history of Heidelberg and the 600th anniversary celebration, and the American students listened dutifully with few opportunities for interaction. Professors on the sidelines and their students yearned for an interruption to start a dialogue, but held to etiquette. In the next two segments, with the Rektor gone, Dr. Must skillfully moderated panel presentations by Dr. Nordheim, several students, and Dr. Must himself. Yet, despite good intentions, he was not quite prepared to draw out the Americans who, without the benefit of a moderator, fended for themselves and did so with admirable vigor.

What we educators had feared before became quite clear during the actual transmission: this event was about promoting the 600th anniversary of Heidelberg University in a transmission that went not only trans-Atlantic, but was also aired live on West German television. This is not what the American students and their professors had bargained for. Understandably, for the Germans, the American students were just voices from a far distance, minor, if somewhat exotic stones in the mosaic of their representation of themselves.

So here was a case of varying expectations. We wanted an educational experiment of pedagogical value. The Germans wanted us to take note of the 600th anniversary celebration. The purpose of the experiment communicated by Intel-Ed to the officials at Heidelberg obviously did not penetrate. And so the desired protocol, which included showing pictures of the American students as they spoke, was not followed.

Post-mortem

For the participating students, the satellite transmission was a hit. They had the feeling of contact with a foreign culture. They enjoyed being part of a telecommunications circus act.

But then we asked ourselves the tough questions. Did the transmissions have any pedagogical value? What is the value, if any, of using satellite video transmissions for point-to-point exchanges? How might telecommunication media be used effectively in language instruction and international education?

Obviously, this experiment does not make a case for the cost-effective use of satellite video conferences in situations which are limited to small numbers of students by virtue of requiring high interactivity. But a case could be made for involving hundreds of classes at many universities in videoconferences whose pedagogical value would not require that students interact with speakers abroad.

Under those circumstances, the transmission from Heidelberg focusing on the 600th anniversary might have been appropriate.

"Little media," such as telephone and radio, in contrast to "big media," such as television, tend to be more appropriate as instructional tools when interactive point-to-point communication is desired. The question is to what extent the visual dimension provided by television is worth the expense.

"Reach Out and Touch Someone"

We think there is a place for international video-conferences involving large numbers of students interested in international cultural and political issues. But for language instruction, or for point-to-point communication among small numbers, we think that the most cost-effective interactive telecommunication medium is the telephone.

We think that for the price of a trans-Atlantic telephone, it is possible to use telecommunications effectively in bringing native speakers to the classroom. There are, of course, preparations that will increase the effective use of this medium. At the minimum, the instructor should have access to a speakerphone and microphone(s) to allow for student participation. Language labs, which are already equipped with microphones and often with the means of controlling microphone inputs, can provide the hardware needed for international telephone encounters. As for the visual dimension, photos and VHS tapes can be exchanged ahead of time. In fact, correspondence should precede such encounters as a way of building up interest and commitment.

Net Gain

Was it worth it? The question has to be posed again. In addition to the substantial work done by the technical teams at Dickinson College and Rutgers University, the language professors put in many extra hours and effort, preparing their students, setting up guidelines, making numerous phone calls to iron out technical details, and attending meetings with Intel-Ed to coordinate the transmissions. There were frustrations, such as having photos of students rushed to Germany only not to be used during the transmission for identification purposes.

Student reactions were positive; even those who were not directly involved and lost class time because of the event, were understanding and supportive. The students who actually took part in the program were absolutely thrilled, and with their adrenalin still flowing, rather verbal in their criticism that too much air time was devoted to lecturing by the German officials.

At Rutgers, the event was carried by the University's own daily, as well as by a number of local newspapers, even the Associated Press. It was a big publicity boost for the German Department, but more importantly for foreign languages in general. Rutgers University's President, Edward Bloustein, who had been an invited guest at the Heidelberg celebrations, told Professor Herbst at a University Senate gathering how his colleague, Rektor zu Putlitz, had been absolutely delighted about the smooth, lively, and cheerful TV-hour linking Heidelberg with America.

But, was it *really* worth it? According to Intel-Ed, it cost about $30,000 to organize and implement this transmission, including in-kind contributions on both sides of the Atlantic. This means that for thirty students it cost approximately $1,000 per student. And for that money, there was very little involvement. Only the more aggressive students jumped at the chance to answer more than once or pose questions themselves. Theirs were precious words, indeed!

In an amicable telephone conversation before the actual broadcast, Dr. Must exclaimed, "So much money! You could easily send all your participants over for that price and we would gladly host them and show them around!"

We have been told that Intel-Ed will continue its experiments with satellite transmissions in pedagogical areas other than language instruction. Meanwhile, we welcome sponsorship that would enable us to take up Dr. Must's offer. Is there anybody who would like to fund such a mundane, old-fashioned, low-tech enterprise? □

KITES: A Middle School Environmental Science Telelink to West Germany

John LeBaron and
Virginia Teichmann

A National Call for Improvements in Global Education

The need for improved global studies in American schools is gaining much attention. In an economically inter-dependent and environmentally vulnerable world, traditional American isolationism seems increasingly inappropriate for effective participation in the global community.

This need is producing a proliferation of school technology projects, some operating from individual buildings and some as national ventures. In 1988, for example, the U.S. Public Broadcasting Service (PBS) conducted a nationally-broadcast 90-minute live video teleconference connecting carefully-selected American high schoolers with their peers in the Soviet Union. In another example, focusing on the poor relative performance of American children on tests of geographical knowledge, the National Geographic Society has been supporting the development of state networks to support the re-birth of K-12 study in geography.

These projects address a variety of themes, including cross-cultural understanding, geography, environmental science, and world peace. Strategies for establishing the teacher/student exchanges range from "low tech" pen pal projects to "high tech" electronic information exchanges. The "high tech" solutions tend to be expensive and complex, making it difficult for ordinary schools to benefit from the learning potential of international communication.

John LeBaron is Associate Professor, College of Education, University of Lowell, Lowell, Massachusetts. Virginia Teichmann is at the Paedogogische Hochschule Karlsruhe, Karlsruhe, West Germany.

Global Studies and Environmental Science: A Rationale for International Telecommunications

Since 1980, a long stream of national reports has warned of a crisis in elementary and secondary science education. This crisis is partly attributable to the difficulty in keeping up with the recent knowledge explosion in science, and partly to inadequate curricular responses to research in science education. Another battery of reports signals a need to improve school productivity and instruction through technology. The University of Lowell felt that its own institutional resources should be marshalled to address these problems.

In a cooperative effort with Digital Equipment Corporation, the Massachusetts Corporation for Educational Telecommunications, and three local middle schools, Lowell has just completed the first phase of a project—called KITES*—to address these problems. In 1988-89, middle schools in three northeastern Massachusetts communities were linked with a West German "sister" school. Using domestic and international two-way television and electronic data conferencing technologies, teachers and students on both sides of the Atlantic shared information on phenomena that affect children's everyday lives. Completing the partnership in Europe was the Paedogogische Hochschule Karlsruhe, a teacher training college in the West German state of Baden-Wuerttemberg, and the Nebenius Realschule, an associated secondary/middle school, also in Karlsruhe.

The American side of this enterprise chose the theme of environmental science as its curricular core. Lowell's fully interactive, seven-town instructional video network (IVN) was used as a training ground for participating teachers and students to become

*KITES: Kids Interactive Telecommunications Experience by Satellite.

familiar with the use of two-way television for learning and instruction. The IVN allowed for the simultaneous live connection of the three participating middle schools to the University and to one another.

The Massachusetts KITES schools were located in the City of Lowell and the towns of Chelmsford and Dracut. Throughout 1988-89, three eighth-grade classes conducted bi-weekly IVN classes on "science, technology and society" (STS) themes. The intervening conventional classwork reinforced and prepared for the IVN classes. Using the human and material resources of the University, KITES classes focused on the scientific principles of nuclear power and alternative energy.

Supporting the two-way television activity was a University-operated electronic conferencing system that encouraged all KITES teachers and project personnel to plan curriculum, share resources, and otherwise exchange relevant information. (This conferencing system resides on one of the University's VAX minicomputers, and is called CoSy—Conferencing System). CoSy is a powerful system permitting creation of open and closed interest-specific conferences, simultaneous electronic conversations, and private electronic mail. A closed conference was established for KITES. All KITES teachers and staff were given CoSy accounts at no charge.

Use of CoSy has exceeded all expectations. Primary conference users were the four project classroom teachers and three University resource people. All were trained to use CoSy's essential features. In the roughly 100 school days since the KITES conference went into operation, approximately 420 individual curriculum planning messages were posted. Not including uncountable private e-mail messages, each professional project participant used the KITES conference, on average, three times a week.

Impediments to the schools' use of electronic conferencing were overcome.

School administrators assured the KITES teachers access to the telephone lines required for regular CoSy utilization. Because of the close proximity of the University and the three participating schools, there were no long-distance telephone charges. Additionally, Digital made ten computer terminals with modems available for home use to the individual project personnel. Three additional microcomputers were loaned to the participating schools. Such access to technology is absolutely necessary to assure that teachers will use it regularly.

International Collaboration: Finding Common Ground

Development on the German side took a different tack. Soon after KITES was first conceptualized at the University of Lowell, an approach was made to Virginia Teichmann, a member of the Paedogogische Hochschule Karlsruhe faculty, who adopted the idea and then selected a school partner. Because one of the American project objectives was to serve a broad range of students, including linguistic minorities and children otherwise at risk of failure, Dr. Teichmann sought a "realschule" (a middle school for pupils not expected to go on to university) rather than a "gymnasium" (university preparatory) as partners for the American students. The Nebenius Realschule was chosen.

Matching the student groups in the two countries proved difficult, owing partly to the differences in educational systems. Because American high school schedules tend to be inflexible, the more loosely-scheduled middle school eighth grade classes (ages 13-14) were sought. The German educators chose a 9th grade class of 16-year-olds (one year away from school graduation). The age discrepancy led to unrealistic expectations, especially on the part of the more mature German pupils.

There were other important differences between the two groups. With one or two exceptions, the American students spoke virtually no German; their German counterparts were substantially more proficient in English. Since the American group connected three classrooms in three different communities, the project involved seventy-five different students, some of whom came from severely disadvantaged backgrounds. The German group was smaller and somewhat more homogeneous.

The German curriculum needs differed from those of their American partners. On the German side, KITES supported instruction in English. For the international exchange, this created problems and opportunities. For example, written and verbal exchanges were undertaken in English, thus eliminating the need for formal translation. However, the exclusive use of English required sensitive handling to deal with the potential German embarrassment of communicating in a foreign tongue and the American embarrassment of *not being able to* communicate in German.

While the size and homogeniety of the German group offered a more manageable project environment, the Americans enjoyed access to a broader range of technological resources, notably the Instructional Video Network, the University minicomputer, and the equipment loaned by Digital. Notwithstanding the relative lack of technology, Dr. Teichmann was remarkably successful in generating private sources of German support. She also secured formal commitments of support from her two state ministries of Education.

Creating Transatlantic Telecommunications Pathways

With the support of the Digital Video Network, KITES teachers and students prepared for a full-motion, 90-minute international video teleconference. This event occurred at 9:00 A.M. (EDT) on June 1st, 1989, providing a live link between the Massachusetts KITES students and their West German partners. Leading up to this event, the participating schools exchanged several mailings of student-prepared "artifacts," including posters, maps, slides, videos, and viewbooks. These artifacts were not particularly related to the theme of environmental science, but rather focused on such themes as hobbies and recreation, fashion, perceptions of each other's cultures, community characteristics, and personal tastes. The reason for this is that the KITES students needed an informal framework for getting to know one another.

International e-mail was handled through the CS.NET worldwide academic electronic data network. American access to CS.NET was provided at no cost to KITES by the University of Lowell Academic Computing Center and the National Science Foundation-supported University Corporation for Atmospheric Research. CS.NET was an essential element of cross-national project management. Without it, the enormous demands of project coordination on two continents could never have been managed. Moreover, CS.NET supplemented the KITES artifact-exchange as a means for project teachers and students to get to know one another.

Although Digital secured and organized all technical resources related to the videoconference, other private companies were involved. Videostar Connections, Inc., a Digital vendor, coordinated the complex arrangements between American and West German telecommunications authorities. International satellite-based telecommunications are regulated by a nearly unfathomable mix of national and international bureaucracies. Videostar helped to cut through the miles of red tape.

The simultaneous two-way live video telecast was carried on PanAmSat, a recently-launched satellite with a "footprint" covering Western Europe and the continental U.S. The June 1st event was the first transatlantic videoconference transmitted via a single satellite. Due to the smaller footprints of older satellites, previous international videoconferences and telecasts have necessitated the use of several domestic and international satellites, involving substantially greater technical and regulatory complexities.

Future Hopes for KITES

KITES has a promising future, but that future depends on higher levels of support. Without the dedication of the University, the three currently-participating schools, the West German partners, and Digital, the project could not have come this far. Without improved support, not only from Digital, but also from other public and private sources, it can go no further. Nevertheless, the project is approaching local corporations and will urge Digital to elevate its support.

If KITES succeeds in securing additional support, several international video teleconferences are planned for the next two years. Future videoconferences may be conventionally broadcast statewide. Massachusetts Educaional Television (MET), the state's school instructional television service, would provide the school-day broadcast time for this purpose. For these telecasts, toll-free incoming telephone access will be provided in order to broaden the range of participating schools beyond the relatively small on-camera group.

For future international videoconferences, prominent leaders in various disciplines will be invited to interact with the teacher-student participants. As the project grows, the growing international community of teachers and students will be encouraged to continue their interactions through international e-mail, data conferencing, and advanced computer file transfers. The project will develop a support mechanism to encourage continued involvement of schools either directly or indirectly affected by KITES. This may be done through the establishment of project-specific electronic bulletin boards, printed and electronic curriculum materials, newsletters, seminars, and consulting opportunities for 'veterans' of previous KITES teleconferences.

Technologies for the Future of KITES

KITES will continue to use a combination of technologies selected for ease of teacher use. Computer and video-based, these technologies will stress the use of telecommunications. The project expects to continue using the CS.NET system for international electronic mail. In those countries not reached by CS.NET, alternative telecommunications strategies will be sought through other service vendors. The project will continue to seek private support for whatever is needed to create the necessary data links.

The international video links present a more complex set of alternatives. The advantage of the live, full-motion analog video technology provided by Digital is that it not only lends itself to simultaneous broadcast over conventional domestic television stations, it also permits the integration of pre-recorded video segments into the live teleconference. Such a capacity would allow for a much broader population of participants than the relatively

small group of schools directly involved in any particular teleconference. The disadvantage lies in the technology's relatively high cost.

Under the lead taken by the Massachusetts Corporation for Educational Telecommunications, KITES has also discussed with several American telephone companies the possibility of conducting test videoconferences using a constantly-improving "compressed video" technology. Although this is a "slow scan" process, the higher grades of compressed video transmit near full motion images. Because high grade telephone lines are used, compressed video transmissions are much less expensive than full-motion analog video. Unfortunately, compressed video fails to meet conventional television broadcast standards.

For the purposes of this project, full-motion and compressed video technologies both appear viable. Because of the conventional broadcast possibilities, the full-motion analog option is preferred. However, compressed video has significant potential for conducting low-cost "practice" videoconferences in preparation for the more complex full-motion teleconferences designed for simultaneous statewide broadcast.

Project Evaluation: Possibilities and Problems

Evaluation results show that students, teachers, project staff, and corporate personnel were pleased with the results of KITES. However, the path from project start-up to the June 1st teleconference was bumpy. The most serious problem centered on the lack of resources for curriculum development and training. Although the in-kind value of DEC's contribution, combined with personnel time from the University of Lowell and other participating organizations exceeded $100,000, there were virtually no funds for teacher release time, for materials, or for consultant help. Such support is critical to long-term success.

Other problems also had to be addressed. In attempting to link the classrooms of three geographically and demographically separate Massachusetts towns, in addition to establishing relationships with overseas institutions and with the private sector, the project became almost unmanageable. In

fact, with resources sufficient for one full-fledged telecommunications project, KITES actually operated as two projects—one domestic, and one international. Future efforts must match expectations to institutional limitations.

The professional cultures of schools and corporations are difficult to bridge. In the case of KITES, the determination to succeed on both sides of the private-public cultural divide combined with the essential good will of the players to overcome tensions. Several adult participants additionally felt that the videoconference itself was "too scripted" and adult-directed, although the students themselves did not cite this as a problem. Since one of the project goals is student empowerment, this perception needs close attention in the future.

Final Remarks

Although KITES has concentrated on environmental science and English language instruction, the thematic potential is much greater. By letting the teachers plan freely via electronic conferencing and encouraging student communication through the exchange of artifacts, videotapes and teleconferences, a broad range of cross-cultural learning occurs. For teachers who routinely need written permission simply to make local telephone calls, and children whose view beyond their city block is defined by commercial television, purposeful international telecommunication can be an extremely empowering experience.

International telecommunication for schools cannot occur without collaboration among diverse organizations. A key to successful collaboration lies in mutual recognition and respect for each partner's fundamental responsibilities, and a clear view of how collaboration will be mutually helpful.

Throughout all of this, the cooperating organizations must communicate, communicate, and communicate some more. In the case of KITES, communication was made all the more complex by the fact that it proceeded simultaneously in more than one country. Electronic telecommunication vehicles helped immeasurably, but the substance and the process came from individuals dedicated to the enrichment of school life by international telecommunication. □

Critical Barriers to the Adoption of Instructional Television in Higher Education

F.R. (Bud) Koontz

Despite the very dramatic developments made in new technologies, it is very evident that there is a cultural lag in teaching methodologies. Colleges and universities have yet to fully adopt the tremendous potential of instructional television (ITV) due to barriers to the innovation itself, faculty career stage and myth barriers, and administrative barriers of budgetary limitations and insufficient training.

The impetus for change, that of adopting instructional television to alleviate the teacher shortage of the late 40s and 50s, is no longer a reason for adopting this technology. The entrenched behavior of academia does not recognize course improvement as using instructional television in the classroom or in a distance learning program. Generally, it also does not reward its users with equal merit pay or promotions. It is human nature, once comfortable patterns of behavior have been established, most especially in academia, that change creates uncertainty, and uncertainty creates frustration, conflict, and even emotional pain. Change also demands additional energy and work. Yet without change in the teaching/learning process, progress in higher education cannot and will not be made.

The study of barriers to the adoption of ITV in higher education, for the classroom and distance learning programs, is *not* a dead issue. First, it is apparent that the wealth of information contained in the literature and research concerning the advantages, benefits and potential of ITV, has been ignored and not adopted or taken seriously. This is evidenced by the extremely low percentage of credit telecourses offered at a given institution compared with the traditionally instructed courses. Second, there are a number of ITV departments which have suffered from the lack of budgetary and administrative support, the result of which

F.R. (Bud) Koontz is Professor of Educational Technology, the University of Toledo, Toledo, Ohio.

caused a reduction in services, personnel, and capital improvements. What was once a thriving ITV department is now history. Or, perhaps, the department experienced a dramatic change in direction and the personnel and equipment are being used for purposes of promoting the college or university. Third, there is now a generation of faculty who have completed graduate school during a time when the demand for the utilization of ITV was low. It is appropriate to review the barriers to adopting ITV, as reported in the 60s and 70s, so that contemporary educational technologists may benefit from what has happened to ITV as they deal with the innovation of the computer in the classroom (Rockman, 1985).

Causes of Barriers

There are five attributes or characteristics that directly affect the rate of adoption of an innovation (Rogers, 1983). These attributes are: (1) relative *advantage* of the innovation, (2) *compatibility* with existing values, previous experiences and needs of the adoptors, (3) *complexity* of use, (4) *trialability* of use, and (5) *observability* of results.

The relative advantage of any innovation is the degree to which the innovation is perceived as being better than the idea or process it replaces. The degree of advantage may be expressed as increased productivity, yielding high economic profitability, or the gain of social status. There are also several subdivisions (Caffarella *et al.*, 1982) of relative advantages such as a decrease in discomfort, saving time and effort, and immediacy of reward. If the innovation is perceived with specific advantages over the present process, it is more likely that the innovation will be adopted.

Compatibility may be defined as the degree to which an innovation is perceived as consistent with existing values, previous experiences, and the needs of the adoptors. The innovation must be compatible with deeply imbedded cultural values and with previously adopted ideas. Old ideas are the main tools with which new ideas are assessed. A potential adopter cannot evaluate an innovation except on the basis of what is familiar. Previous and present practice is a standard against which the innovation can be judged, thereby decreasing uncertainty. Potential adoptors may be unaware of their own needs for the innovation until the innovation and consequences of adoption are manifested. When adoptor needs are met, a faster rate of adoption usually occurs.

Complexity is the degree to which the innovation is perceived as being relatively difficult to understand and use. Innovations can even be placed on some type of simplicity-complexity scale. Some innovations are easy to understand while others may require specialized training over an extended period of time, the assistance of a consultant, or the hiring a full time professional. Therefore, the more complex an innovation becomes, as perceived by the potential adoptors, the lower the rate of adoption.

Trialability is the degree that an innovation may be used on a limited basis. Innovations that may be tried for a short and specified period of time with a minimum investment of time and finances tend to have a greater adoption rate than those innovations that require a longer period of time with a higher investment. Trialability reduces risk and allows the reversion effect to occur, i.e., return to the status quo if the innovation does not prove satisfactory.

Observability is the degree that others may see the result of the use of the innovation. The more positive the observed results, the greater likelihood that the innovation will be adopted. However, some results are more observable than others, thereby increasing the adoption of the innovation.

Teaching faculty, who are the potential adoptors, unknowingly create barriers by their own traditional way of evaluation for salary, promotion, and tenure. There are three specific *career stages* in which faculty may be categorized (Wedman and Strathe, 1985). These stages are the early entry and career stabilization stage, the midlife development stage, and the senior life stage. The early entry and career stabilization stage is characterized by the faculty member seeking to secure a position within the institution. This is accomplished by establishing professionalism in teaching, service to department, college and institution, publishing, and conducting research. The midlife development stage is characterized by reassessing personal and professional lives. This stage may be characterized by professional withdrawal and enthusiasm for change in the manner in which one teaches or is evaluated. Personal concerns and needs may be more important than institutional needs. The senior life stage involves the faculty member preparing for retirement and who has for the most part resolved the issues of midlife development. Faculty often return to active participation and support of institutional change. They may want to share with the early entry and career stabilization group lessons that have been learned. The needs and concerns of senior faculty are primarily centered on how they can meaningfully continue to participate in the institution.

Few research studies and articles have been published concerning the faculty attitude toward the use of instructional television. Evans (1968) in *Resistance to Innovation in Higher Education* and Dubin and Hedley (1969) in *The Medium May Be Related to the Message: College Instruction by TV,* both concluded that the paramount reason for faculty not using television for instruction is the belief that good teaching depends upon direct contact of faculty with the students. Evans also concluded that professors prefer using familiar teaching methods and that any teaching method that is innovative is met with anxiety and apprehension.

Faculty and administrators have also developed a number of misperceptions and myths concerning the adoption of ITV which have become rigid barriers (Luskin, 1983). Examples of misperceptions include that television courses only require viewing and are relatively easy when compared to the *traditional* courses. There are only a few quality telecourses that could be integrated into a program of study. And further, telecourse students do not learn as much as traditional students due to the shallowness of course content. There are a number of myths that faculty have developed which include: (1) faculty have no specific role in the teaching function when a telecourse is offered; (2) the technology of producing a television recording of instruction is somewhat of a fad and will be disregarded as a method of teaching; (3) the use of instructional television dehumanizes the teaching-learning experience; (4) telecourses create an inflexible form of instruction since the pace of the television lecture is fixed; (5) there is a lack of evidence that supports the notion that students learn by television; (6) students have no immediate feedback during the televised lecture and testing; (7) there is no personal contact with the professor who is teaching on television; and (8) the student lacks the necessary motivation to do well in the course. The list of myths appears to be virtually endless.

In addition to the "built-in" barriers to the adoption of the innovation of ITV in higher education and barriers that the potential adoptors possess, there are specific administrative barriers. These barriers include the lack of budget allocations, the lack of understanding and motivation by administrators, and the lack of training for the professor and administrator in using instructional television in higher education.

Budget allocations for the purchase of television equipment has been shrinking. Many of the outside funding sources, once available from the government, no longer exist. The original hardware purchased for university ITV productions is inadequate, obsolete, and in many cases not replaced when a complete breakdown occurs. Insufficient funding does not permit release time for professors to prepare an ITV series. No funding is available to financially reward the professor for spending additional time in producing an effective ITV course. When a series is produced, however, the professor is faced with the problem of ownership if the series is to be marketed to a professional distribution agency. There is insufficient funding for ITV program development, production time, or supplies for the studio, or the hiring of a qualified ITV producer/director.

It also appears that there is generally a lack of administrative encouragement for the professor to become involved in an ITV series or project due to the extended time required for writing and producing. The additional time spent with ITV subtracts from the time the professor would ordinarily spend in class preparation, student contact, and other professional duties. When faculty are evaluated by administration, the production of an ITV program or series is not considered equivalent to a published journal article or book. With little or no recognition given, the professor must revert to the traditional norms of acceptance and recognition and publish in the traditional formats to be eligible for a merit increase. Negative feedback is also quite possible. Exposing one's teaching deficiencies before a peer group and administrators opens up the faculty member to even greater risks!

While there is a lack of training for educators and administrators to use a technology of instruction, professors present a different type of problem to the educational system. College professors are required to be knowledgeable experts in their field of study but are not required to demonstrate any actual teaching skills. The majority of faculty have identified, at some time during their course of study, a professor who has been admired, and adopted much of that teaching style. Administrators, for the most part, have been teaching faculty who have exhibited management skills in such a way as to receive an academic promotion that takes them out of the classroom and places them into an office where the problems and importance of teaching are soon forgotten.

Effects of Barriers

The rapid growth spurts of the late 50s and 60s and the availability of the National Defense Education Act of 1958 to finance many ITV facilities appeared to indicate that this innovation was well on its way to being adopted by nearly all institutions as an integral part of their educational system. By 1983 (Luskin) approximately 800 or one-third of the colleges and universities regularly offered ITV courses. The Adult Learning Services (ALS) of PBS, in its 1987 report, *The First Five Years*, indicated that almost one-third of the nation's colleges and universities have offered ALS ITV courses to approximately 500,000 students.

However, other reports concerning the use of ITV in higher education are less enthusiastic. Only about twenty percent (Hillard, 1985) of American schools use television on a regular basis (this includes elementary through higher education) and only 17 percent of the teachers have had any training in the purpose or utilization of ITV. When ITV courses are compared to the conventionally taught courses at institutions that have adopted ITV, the percentage is exceptionally low.

Some ITV apologists, however, maintain that ITV has never been accepted and that it has never obtained a solid academic position in higher education. This might have been due, in part, to the fact that television has not delivered the educational gains it promised. Part of this disenchantment may have been due to the result of copious research studies that did not clearly indicate that ITV was a superior way to teach. Conservative research designs of providing the same information and teaching style by the professor in both the traditional classroom and televised lectures did not truly test the medium of television. The inability of television to manifest any instructional effectiveness prevented whatever chance there might have been for it to have become an integral instructional component.

It appears that the advantages of using ITV also have diminished. The impetus to use television to solve the teacher shortage no longer exists. Due to the complexity of use, which now requires the hiring of a professional to produce and direct the ITV program and technicians to maintain the system, faculty are then intimidated to use the technology. The trialability of use allowed the door for the traditional teaching style always to remain open. The reversion effect was bound to happen with no real commitment to the use of television teaching. When the compatibility of using ITV clashed with existing values of teaching in front of the classroom, it was yet another blow to adoption. Research did not provide the results required to encourage adoption. The conservative research approach became entwined with the covert politics of not proving that television perhaps was a better way to teach. This, to some extent, was done to perhaps hush the faculty cries of fear that television would replace them as classroom teachers.

It also appears that faculty career stages do not allow for any adoption to occur, either. For faculty in the early entry and career stabilization stage, the importance of obtaining tenure dramatically outweighs the experimentation with a questionable and little researched teaching innovation. The midlife stage faculty do not accept the thought of changing their style of teaching in front of a classroom that has taken several years to develop and has rewarded them with the necessary tenure and promotions. The senior life stage faculty is no different. The senior professor realizes that a high merit increase closer to retirement is more important than a change of teaching style.

Results

What direction should be taken at this point to increase the use of ITV in higher education? Are the barriers too large and numerous to overcome? What consolation or comforting words might be given to soften the disappointment and discouragement of those who have dedicated their entire professional teaching careers to the use of instructional television? Should the use be abandoned, modified or should the present adoptors become better change agents and thereby more effectively promoting the continuance? These issues would require another paper of equal length. However, there are some creditable suggestions that would be effective persuasion techniques to increase the adoption of ITV in higher education and to make the technology more effective. These suggestions are: (1) create needs for the adoption of ITV; (2) exemplify the advantages of the innovation; (3) demonstrate how the innovation is compatible with the existing teaching culture; (4) communicate the successfulness of its use and where it has been used successfully; (5) debunk the many misunderstandings and myths about ITV; (6) create diffusion models that will create greater adoptability; (7) perform a needs assessment; (8) develop support groups of faculty adoptors that will advocate its use and assist in the training process.

The technology of instructional television is dynamic, it is improving, and hopefully it is here to stay. Universities and colleges will never be able to claim that they really care about the learning process until professors and administrators are required to exhibit knowledge and proficiency in the teaching process, including the ability to use instructional television. ☐

References

Books

Dubin, R., and Hedley, R.A. *The Medium May Be Related to the Message: College Instruction by TV.* Eugene: University of Oregan Press, 1969.

Evans, R.L. *Resistance to Innovation in Higher Education.* San Francisco: Jossey-Bass, Inc., 1967.

Gordon, G. *Educational Television.* New York: The Center for Applied Research in Education, Inc., 1965.

Hillard, R.I. *Television and Adult Education.* Cambridge: Schenkam Books, Inc., 1985.

Rogers, E. *Diffusion of Innovations* (3rd ed.). New York: The Free Press, 1983.

Journal Articles

Caffarella, E.P., Caffarella, R.S., Hart, A.E., Poller, A.E., and Salesi, R.A. Predicting the Diffusability of Educational Innovations. *Educational Technology,* 1982, *22*(12), 16-18.

Duttweiler, P.C. Barriers to Optimum Use of Educational Technology. *Educational Technology,* 1983, *23*(11), 37-39.

Hendrick, L.C. Strategies to Encourage Greater Faculty Use of Instructional Television. *Community College Review,* 1986, *14*(2), 27-31.

Luskin, B.J. Telecourses: 20 myths, 21 realities. *Community and Junior College Journal,* 1983, *53*, 48-52, 60.

Rockman, S. Success or Failure for Computers in Schools? Some Lessons from Instructional Television. *Educational Technology,* 1985, *25*(1), 48-50.

Rose, S.N. Barriers to the Use of Educational Technologies and Recommendations to Promote and Increase Their Use. *Educational Technology,* 1982 *22*(12), 12-15.

Wedman, J., and Strathe, M. Barriers to Optimum Use of Educational Technology. *Educational Technology,* 1983, *23*(2), 15-19.

The Use of Interactive Television in Business Education

Gordon D. Pirrong and William C. Lathen

Faculty in the university environment need to find more effective and efficient methods of program delivery. In an article by Decker and Krajewski (1985), it was stated:

> What educators need is a better understanding of the opportunities that new technology brings to the education profession. Teachers need to realize that technological advances can enhance/improve the quality of instruction.

Educational literature has been critical of "TV" classrooms because the students could not communicate with the instructor and the dialogue was only one-way.

Interactive Television Classrooms

Recent technological changes have given students the ability to communicate with their instructors during class, using interactive television classrooms. The interactive television classroom allows the students to hear, see, and to *communicate* directly with the instructor. Lundgren (1985), however, is skeptical of using an interactive television system for the delivery of educational programs:

> Two-way interactive systems are not expected to be as good as, or better than, having the students in the actual classroom with their teacher.

Koontz (1989) has reported that many faculty and administrators have developed a number of misconceptions which have become barriers to the adoption of TV courses.

In a survey of three high schools that used an interactive instructional television system, Nelson (1989) reported that the students perceived little difference in the interactive television class and the traditional classroom. In addition, the teachers involved reported *no significant differences* in the 'students' test scores, grades, and participation when comparing sections of TV classes and classes taught in the traditional classroom.

Gordon D. Pirrong and William C. Lathen are Associate Professors of Accounting at Boise State University, Boise, Idaho.

A review of the literature shows that there are conflicting opinions about the usefulness and effectiveness of interactive television in the general classroom. None of the studies focused on business education or classes at the university level. This study examines the effectiveness of using an interactive television system to broadcast an introductory accounting class at the college level.

Study Design

To test the use of the interactive television program delivery method, one section of an introductory financial accounting course was broadcast from Boise State University (Idaho) to three remote sites with 16 students. In addition, 34 students were also present in the TV classroom during the broadcasts. Another section of the course was offered to 21 students on the same days but in a traditional campus classroom.

The study was designed to test the general hypothesis that student performance and attitudes should not vary significantly whether the student was in the traditional classroom, the on-campus interactive television classroom, or the remote site interactive television classrooms. All three classes (traditional classroom, on-campus ITV classroom, and remote site ITV classrooms) were taught by the same instructor and met the same days for 50-minute sessions, three times a week.

The performance of each of the three student subgroups were evaluated on identical scales:

5 Quizzes	50 points
4 Exams	400 points
Term Project	20 points
Homework	30 points

Exams and quizzes at the remote sites were administered by special proctors at the same time as in the ITV classroom. Thus, no group of students had a time advantage over another group. Exams and quizzes were multiple-choice and problems. All were graded by the instructor.

Two questionnaires to measure student attitudes were administered during the last week of the semester. The questions were separated into five categories that measured attitudes about the (1) course, (2) instruction, (3) instructor, (4) facilities, and (5) student demographics.

All of the opinion questions used a five-point Likert scale (STRONGLY AGREE, AGREE, NEUTRAL, DISAGREE and STRONGLY DISAGREE) and a NO OPINION CATEGORY. The questions were designed to measure the students' perception of the quality of their educational experience.

Table 1

Student Scores—500 Possible Points

Location of Student	N	Mean Scores	Std. Dev.
ITV Remote Sites	16	395	52
ITV On-Campus	34	374	48
Traditional, On-Campus	21	364	38

Table 2

ANOVA on Student Performance

Source	df	Mean Square	F Value	P Value
Model	2	4485	2.02	0.1407
Error	68	2222		

Study Results

The performance results of the three groups of students are summarized in Table 1. Notice that although the ITV remote site students scored the highest average grades, they also had the largest standard deviation. Conversely, the traditional students scored the lowest overall average of the three groups, but they also had the smallest standard deviation.

Performance Results. The first hypothesis states that no significant difference should exist in the students' mean performance scores among the three different classroom settings. Table 2 presents the results of the One-Way ANOVA on students scores for the three groups.

The P value (significance level) of 0.1407 is sufficiently high so that the first hypothesis cannot be rejected. This result supports the argument that average performance scores are not significantly different between students taught via ITV to off-campus sites and students in the traditional on-campus classrooms.

Attitudes. The second hypothesis asserts that no significant difference exists in attitudes about the class experience whether the student was in the traditional, the on-campus ITV, or the remote ITV classroom setting. The Smirnov test for differences is appropriate when testing for differences between subgroups when the measurement scale is ordinal (Gibbons, 1976).

For several of the questions asked the students, for example, question 30, about location, it was expected that the remote site classrooms would have a lower average response than the traditional or on-campus ITV classroom. For other questions, for example question 13, about age, the expectation was that the remote site would have a higher average response. For these types of questions, since their general characteristic is known, the more powerful one-sided Smirnov test is proper because differences in attitudes are more easily identified than when using the two-sided test. Table 3 depicts the results of the questions that display potential differences.

Differences in Attitudes. The results of the Smirnov test are interesting. First, teaching ability questions, such as instructor preparation, knowledge of the subject, and overall teaching ability showed *no significant differences* among the three groups. The students from the three classes evaluated the instructor the *same* regardless of where they received the instruction.

The remote site ITV classes, however, did show significant differences in satisfaction for three questions, 20, 21, and 22. All three of the questions indicate a significantly higher level of dissatisfaction by the remote site group. For question 20, concerning the ability to communicate with the instructor, evidently the ITV system caused some problems. The problem may have been as simple as the reluctance of the students at the remote sites to use a microphone to talk to the instructor. The negative feelings could also be caused by non-classroom communication problems, such as the student not being able to talk to the instructor before

Table 3

Smirnov Test for Differences Between Remote Site Classrooms vs. Traditional or On-Campus ITV Classrooms*

Question #	Question Description	One-Sided or Two-Sided Test		Difference between Remote Site and	
				Trad. Class	On-Campus ITV Class
13	Age	> one-sided		.01	.05
14	Class standing	two-sided		.05	.10
15	Course required or elective	> one-sided		.01	.01
18	Gender	two-sided		.01	.02
20	Able to communicate with instructor during the course	> one-sided		.05	.10
21	Instructor's writing was legible	> one-sided		.10	.05
22	Instructor's visual aids readable	> one-sided		.05	.05
30	Location a main reason for taking course	< one-sided		.01	.01
34	Student would take another ITV class	< one-sided		N/A	.41
36	ITV made it easier to take the class	< one-sided		N/A	.01
37	ITV interfered with learning	two-sided		N/A	.40

* Only significant results of .05 or lower are reported, plus questions 34 and 37.

< Remote site expected to be closer to 1 than other groups.
> Remote site expected to be farther from 1 than other groups.

or after the class, or the student not able to come to campus during the instructor's office hours. Many of these problems are inherent to an ITV system, and may not be correctable.

The remote site students also expressed dissatis-faction in answering question 21 (instructor's writing was legible) and question 22 (visual aids were readable). Since there was generally less problem with this in the more traditional classrooms, perhaps the camera did not hold the picture long

Table 4

Student Attitudes about the ITV System

Question # & Description	Type of Response	ITV Students	
		Remote Site	On-Campus
34 - I would take another ITV course.	Strongly Agree	22%	29%
	Agree	50	38
	Neutral	21	12
	Disagree	7	21
	Strongly Disagree	0	0
37 - ITV, as used in this course, interfered with my learning.	Strongly Agree	7%	4%
	Agree	6	15
	Neutral	47	23
	Disagree	27	35
	Strongly Disagree	13	23

enough for the student to completely absorb the material presented or perhaps there were some camera focusing problems. Those troubles are correctable, but the problem may be more severe. For example, the ITV student dissatisfaction may be caused by the camera focusing on the chalkboard or visual aid adequately, but the student is unable to see the instructor continually, as do students who are in the same room as the professor.

Question 30, "One of the main reasons I enrolled in this course is because it was offered at this location," was significantly different for the off-campus group compared to the on-campus groups. It is evident that the remote site ITV course attracted students that otherwise would not have come to campus to take the course.

Questions 34, 36 and 37 provided the best information concerning the attitudes of remote site ITV students. Question 36, "The use of ITV made it easier for me to take this course," was significant at the 99% confidence level. Evidently, the off-campus students are satisfied with the remote site classes. Question 34 reinforces this position. Question 34, "I would take another ITV course," showed no significant difference between remote site and on-campus ITV students. Neither group displayed significant dissatisfaction with the ITV system. Further, for question 37, "ITV . . . interfered with my learning," there was no significant difference between the two ITV groups.

The percentage of student responses for question 34 and question 37 are shown in Table 4. The results of question 34 show that 93% of the remote site ITV students and 79% of the campus ITV students would take another ITV class. The responses to question 37 indicate the majority of off-campus students do not believe their learning experience was impaired by using the ITV system. Table 4 does, however, indicate a small segment of dissatisfaction with the on-campus ITV students. In this group, 21% disagreed with question 34, "I would take another ITV course," while 19% either agreed

Table 5

Demographic Characteristics

Question # & Description		Traditional Classroom	ITV Students	
			Remote Site	On-Campus
13 - Age	Under 26	100%	50%	81%
	26 or over`	0	50	19
14 - Class	Lower division	95%	93%	72%
	Upper division	5	7	28
15 - Course is	Required	95%	36%	96%
	Elective	5	64	4
		100	100	100
18 - Gender	Male	74%	20%	79%
	Female	26	80	21

or strongly agreed that ITV interfered with their learning (question 37). Although there were no significant differences in grade performances, a number of the on-campus ITV students may have felt hindered because of class time delays inherent with the ITV system.

There are five demographic questions and four of them have significantly different responses among the three groups (questions 13, 14, 15 and 18 from Table 3). These are summarized in Table 5. The demographic questions indicate that students at the remote sites were older, had a lower class standing, were taking the course as an elective and the majority were female. The mean test scores of the three groups showed that the remote site students had the highest average total points but the difference was not significant. This reinforces the conclusion that the ITV delivery system is effective because even with the demographic differ-'ences, no significant differences were found in grade performance or in the majority of the opinion questions. The remote site students are performing at least as well as the on-campus business degree students.

Summary and Conclusions

The purpose of the research project was to determine if significant differences exist in university business students' performance and attitudes when the classroom setting is a remote site ITV classroom compared to the traditional classroom or a classroom in which the ITV broadcast originates. Student performance was evaluated through examinations, quizzes, and homework covering various parts of the course. At the end of the semester, students were given evaluation forms that contained 39 questions about the course, instructor, instruction, demographics, and classroom facilities.

There were no significant differences among the three groups in student performance based upon an overall point tally. There were significant attitude differences about the ability to communicate with the instructor, the instructor's writing legibility, and the readability of visual aids by the remote site ITV students. This dissatisfaction was offset by the perceived benefits from taking the course through ITV, primarily saving time from not having to travel to and from campus. A large majority of the remote site students liked the overall system sufficiently well to take additional ITV courses.

The results of this study support the argument that interactive television is an acceptable method for teaching accounting in a university and should be useful in delivering business education courses. Student performance is not hindered and their attitudes are good. □

References

Decker, R. and Krajewski, R.J. The Role of Technology in Education: High Schools of the Future. *NASSP Bulletin*, November 1985, 2-6.

Gibbons, J. *Nonparametric Methods for Quantitative Analysis*. New York: Holt, Rinehart, and Winston, 1976.

Koontz, F.R. Critical Barriers to the Adoption of Instructional Television in Higher Education. *Educational Technology*, April 1989, 45-48.

Lundgren, R.W. Two-Way Television in Rural Curriculum Development. *NASSP Bulletin*, November 1985, 15-19.

Nelson, R.N. Two-Way Microwave Transmission Consolidates, Improves Education. *NASSP Bulletin*, November 1989, 38-42.

The Independent/Distance Study Course Development Team

Clayton R. Wright

Introduction

There is increasing interest among traditional community colleges in the development of self-paced, independent study courses. These courses may be delivered either on site in learner centers or via distance delivery. These alternative delivery methods are designed to meet the needs of adult students who are restricted by job and family responsibilities or who require retraining. Recently, traditional community colleges have been focusing on the unique requirements of these individual adult learners. Traditional instruction—with its rigid entry points, lockstep sequencing and pacing, as well as inattention to *individual* learning needs—is exploring alternatives that will allow adult learners the flexibility they need to access educational opportunities (Holmberg, 1981). One of these alternatives is independent/distance study; some refer to this alternative as open learning.

Grant MacEwan Community College (GMCC), Edmonton, Alberta, Canada, a traditional institution, is in the process of developing independent study courses. Since most of our classes are delivered in the traditional lecture manner, creating courses for delivery via independent study becomes a major challenge. Why? GMCC instructors traditionally prepare and deliver their courses to a large on-campus group. Independent study courses, however, may employ an instructional design systems approach whereby a course development project team provides input and prepares the necessary course materials. Many of GMCC's traditional instructors have rarely, if ever, worked with such a project team. They are inexperienced at providing course material for their "invisible" distance students and, therefore, find it difficult to prepare materials without receiving instantaneous feedback concerning the content and delivery of their courses. Nor are they familiar with the task of complying with production timelines and accept-

Clayton R. Wright is Coordinator of Instructional Development at Grant MacEwan Community College, Edmonton, Alberta, Canada.

ing the intrusion of non-academics into the teaching process (Kelly, 1987).

In order to address the above concerns, the GMCC Instructional Development Department (IDD) established guidelines for independent study course production. These guidelines are described in Wright (1987), the *Course Developer's Manual*. This article, however, discusses only the roles, responsibilities, and work flow within a course development team. Although the ideas presented here are not innovative in the true sense of the word, they are new to our traditional college. The model presented here for course team composition and operation is neither unique nor ideal; it is only one model that has proved to be successful.

In order to accommodate independent study course production at GMCC, no attempt was made to change the organizational structure or alter the responsibilities of staff at the institution. Instead, IDD tried to develop an independent study design model which would accommodate the existing historic, administrative, and academic responsibilities of the institution.

Rationale for Using a Course Development Team

The importance of employing a course development team approach in independent study or open learning has been stressed by Perry (1976) and others. A team is ideal for completing complex, sophisticated, or large projects such as the development of multi-disciplinary courses or an entire academic program (Kelly and Haag, 1985). Independent study courses produce a public record of instruction (Kelly, 1987), which can be used to judge the quality of an institution's instruction, whereas traditional instruction is primarily transient (Shaw and Taylor, 1984). Consequently, there is a need to produce high-quality courses. Usually, teams can produce a better product than a single instructor can.

The advantages of using a course development team (Moisey and Wright, 1987) include:
- working with experts who are knowledgeable about course content and course production;
- fostering an exchange of ideas;
- developing a collegial spirit;
- taking advantage of the synergy that occurs when two or more people work together to generate a solution to a problem;
- promoting staff development.

The use of a course development team can have the following *disadvantages*:
- instructors may feel that their academic freedom has been curtailed (Perry, 1977);
- teams are difficult to manage; communication and coordination difficulties as well as

personality conflicts may arise;

- large teams require a sufficient student population and financial support to justify their existence (Kelly, 1987);
- team production may be more time-consuming than production by one person.

Solutions to many team management difficulties are addressed in Frame (1987), *Managing Projects in Organizations*.

Despite the drawbacks, a course development team can produce effective and efficient instruction if a systematic, planned process of production is implemented (Smith, 1987). It is advisable to form course development teams in the following circumstances:

- when an entire program is being revised;
- when the course under development is multi-disciplinary;
- when the course is part of a professional program certified by an external organization;
- when an external organization, such as the provincial or federal government is involved;
- when the best possible product is desired.

The size of a course team will vary. A large course team may be assembled to work with writers who are preparing their first independent study courses. As writers become more experienced, the size of the team may decrease. The need for large teams will also decline as faculty are exposed to independent study methods (Kelly, 1987). For example, last year IDD offered 12 workshops on independent study. As a result, faculty gained substantial skills. Consequently, the design team in two programs was reduced to three members each. When instructors take on the production of independent study courses, they tend to accept their new roles as members of a course development team as they accepted their previous roles as traditional instructors. Further, traditional institutions will soon place the same emphasis on producing independent study courses as they do on preparing on-campus courses. When this occurs, the size of course development teams will be reduced.

Composition of the Project Team

Many independent study courses recently developed at GMCC involved a project team. Large projects may involve many people in different roles; in smaller projects, two or three people may perform all the functions. However, responsibility for course production is still divided between instructors and non-academics.

Listed below are suggested course development team members as well as their roles and responsibilities for a large project. Although the composition of such a team will vary from project to project, or from institution to institution, the team performs most of the functions listed below (Wright, 1987).

Academic Dean/Program Head

- may delegate the authority to manage projects;
- assumes ultimate responsibility for the completion of course development projects;
- obtains human and financial resources to mount projects.

In some dual-mode institutions the office of distance or continuing education may perform these duties.

Project Coordinator

- the program head may assume this role;
- acts as the day-to-day project manager;
- is responsible for preparing the project manual that describes the parameters of the project;
- identifies the content to be covered;
- establishes timelines and specific team responsibilities;
- monitors work flow and project progress;
- monitors project expenditures;
- sets an acceptable standard for all work to be completed;
- reviews all materials;
- signs off the final draft before it is printed.

Several distance institutions, particularly those institutions that do not have academics on staff, assign the above roles to the instructional designer. At GMCC, project coordination is the responsibility of each academic area; program heads and deans usually have the final say about academic standards, personnel, and budgets. IDD assists with these activities, however. In a few cases, IDD acts as the project coordinator, but this is usually an informal arrangement.

Resource People

- the members of a program's advisory committee or individuals working in the field may be suitable as resource persons;
- contribute to the overall planning of the project, based on their varied experiences, by describing what has been done, what should be done, and the characteristics of the potential audience;
- prepare the project manual;
- contribute content or resources;
- provide feedback as each course is developed;
- evaluate output of the development process.

In industry/business environments, resource people are often referred to as subject matter experts (SMEs); they primarily contribute content to the development process. In traditional institutions, the course writers perform this function; usually they are instructors who are knowledgeable about the program's content. In both situations, a network of resource people must be developed in order to validate the accuracy and currency of the courses under development.

Instructional Assistant
○ assembles course material;
○ completes work order forms;
○ liaises with course developers and service departments such as the media and photo-reproduction units;
○ obtains copyright releases;
○ distributes course material.

The instructional assistants at GMCC perform a variety of duties that support course development and implementation. Dedicated distance institutions may assign individuals or whole departments to perform the same tasks completed by instructional assistants at GMCC. For instance, at larger institutions, obtaining copyright permission is a task frequently assigned to one or more individuals. At other institutions, the course developer/writer, the instructional designer, or the librarian may assume this responsibility. Often, however, writing for copyright permission and maintaining copyright records is handled by an instructional assistant.

Course Developer/Writer
○ prepares course manual with the assistance of the instructional designer and project coordinator;
○ researches and writes course material;
○ decides on course content and module sequence within a unit;
○ identifies appropriate media;
○ develops assignments and tests;
○ sets evaluation procedures;
○ prepares both student and instructor material related to a particular unit of work;
○ prepares graphics and other media requirements;
○ ensures the fair and honest use of copyright material;
○ proofreads and checks course materials before final production.

The above activities are common to all institutions that develop independent study or distance delivery course materials. Large, dedicated institutions usually require their course developers to perform only those items listed above, whereas smaller institutions or institutions involved in dual-mode education may require their course developers to perform additional functions. At GMCC, individuals who combine knowledge of content, instructional expertise, and writing experience are often employed as course developers. The best

lecturers or management-oriented program heads may not necessarily be the best writers for independent study courses.

Course Reviewer
○ the program head, a faculty member, or a resource person often assumes this role;
○ ensures that the course content meets academic standards and that instructional objectives are met;
○ ensures that course materials are accurate and current;
○ ensures that the content is appropriate for clients who must operate in the "real world."

Course reviewers are exceedingly important when the institution lacks the appropriate content expertise or when a new course or program is being established within the traditional institution. Many distance institutions, who do not have academics on their permanent staff, rely on the course reviewers to verify the academic standards of a course.

Instructional Designer
○ orients the course development team to independent study and the ramifications of on- and off-campus delivery;
○ acts as a consultant to course developers when blueprinting the course and when establishing instructional objectives;
○ aids in rewriting objectives; checks for quality, clarity, and readability;
○ reviews modules for content organization; sequences content, usually in order of ascending difficulty, although chronological, hierarchical, generic, inquiry-related, utilization-related, or spiral sequencing may be employed;
○ reviews evaluation strategies and test items;
○ provides basic editing services;
○ critiques drafts;
○ designs CML/CAI courseware, and trains instructor on CML/CAI systems;
○ provides information on alternative media, distance education, and course management systems;
○ suggests media that will improve the finished product;
○ monitors the progress of the project;
○ reviews all developed course materials;
○ takes part in the formative evaluation of the course materials; checks that students understand instructional concepts;
○ tries to achieve a "win-win" situation with regard to his or her client contacts.

England (1987) noted that instructional designers must be careful not to promote a content-oriented, linear approach to course design. Instead, designers must apply insights from research in the areas of cognitive psychology and artificial intelligence.

These insights suggest that the designer should ensure that:

- learning tasks address the needs of learners who employ different learning styles;
- a flexible learning environment is created;
- students can relate personally to the materials;
- processing techniques and content are equally emphasized;
- students develop a cognitive schema for the material so that later interpretations of the material can be processed at a greater depth (England, 1987, p. 13).

At GMCC, IDD provides all instructional design services.

Editor

- performs substantive editing, copy editing, and proofreading;
- substantively edits: comments on the overall organization of the content; clarifies what the writer wants to say; ensures that the format and style are consistent with the course developer's manual and acceptable style manuals;
- copy edits: checks for correct grammar, punctuation, and spelling;
- proofreads: checks for typographical errors; compares final copy to the original;
- proofreads the final copy before and after paste-up.

In many distance education institutions, the instructional designer is also the editor. At GMCC, however, there are several difficulties with this approach. First, the designer is rarely an expert at *everything*. Substantive editing requires high-level writing skills beyond basic copy editing and proofreading skills. Second, the editor can provide a dispassionate view of the material. Unlike the designer, the editor has no ownership of the material. Finally, when given a choice between having instructional designers or editors, IDD chose to have instructional designers on staff. Designers must be available for consultation during the entire course development process whereas editors are needed only at specific stages, and their services can be contracted as needed. This arrangement is ideal for smaller institutions.

Word Processor Operator

- provides data entry;
- advises on capabilities of word processing system;
- advises on procedures for the most efficient use of system.

Although more course writers are inputting their own text, professional word processor operators are still needed to alter and format course materials.

Media Specialist

- suggests appropriate media to enhance or support learning materials;
- prepares all print materials, including headings, titles, page layouts, graphics, PMTs (copy material for photocopying, enlargements, or reductions), paste-ups, camera-ready copy, laser printing, and binding;
- prepares audiovisual materials such as 35mm slides, photographs, audiocassettes, and videocassettes.

Several media specialists may be involved in developing a set of course materials. For example, a graphic artist may be employed, as well as an audiovisual technician.

Research Assistant

- conducts needs assessments;
- provides audience analysis data;
- designs summative evaluation procedures for course/program implementation;
- conducts graduate follow-up surveys.

IDD or the Research, Development, and Evaluation Department can provide these services. The latter department can provide specific information on the required competencies in a given field. This information is particularly important if the course under development is for a job- or trade-specific program.

The First Project Team Meetings

Once the members of the course development team have been selected, the team will meet several times before the course material is actually produced. Team meetings are necessary if the goal is to produce cohesive and effective course materials in an efficient manner.

It is essential that all members of the course development team attend these meetings. Team meetings:

- establish a common bond among team personnel;
- permit the group to develop a common focus;
- set the standards by which future work will be measured;
- are an effective way for the group to communicate.

Through these meetings, concerns can be identified and frustrations reduced in the early stages of the course development process.

Each team meeting should achieve the goals outlined below.

Meeting 1

- define the project and establish goals;
- identify the educational gap that the course or program will fill;
- set basic project parameters;

○ introduce the project participants and describe the potential contributions of each member;

○ establish the concept of a team effort;

○ introduce the team to independent/learner-centered/distance study.

Meeting 2

○ redefine parameters and constraints outlined in Meeting 1;

○ establish the roles and responsibilities of each team member;

○ identify the characteristics of the potential learners;

○ allow course developers to outline the content they are to write;

○ outline work flow;

○ obtain commitment from participants.

Meeting 3

○ review the detailed outlines for each content area;

○ establish basic writing style, print format, and copyright procedures;

○ establish a definite work flow within a specified timeline;

○ review each person's contribution and responsibilities.

After these three meetings, the entire course development team may not meet again until the finished product is reviewed and revised. Nevertheless, other meetings may be held to:

○ review the course/program blueprint;

○ discuss writing style;

○ develop interpersonal skills;

○ review material as it is produced;

○ discuss organizational or production difficulties;

○ update team members on the progress of the project;

○ receive training on or assistance with particular aspects of course development;

○ introduce team members to media options and distance delivery technologies. Designing a specific hands-on demonstration or exposing team members to a technology fair sponsored by the institution (Wright, Zwicker, and Conroy, 1987) are two ways of introducing team members to instructional alternatives.

Work Flow Within the Team

During the first team meetings, a work flow for the print course materials is established. A typical work flow pattern for a large project is illustrated in Figure 1. Ideally, work progresses according to this pattern whenever a module, student guide, course guide, or instructor's manual is being designed. Note that Figure 1 is actually a copy of a GMCC worksheet that is attached to envelopes

containing written material under development. The sheet is used to record the path of each module as it circulates through the course development process. The appropriate square is signed and dated before the material is passed on.

Conclusion

Using the project team approach described in this article, GMCC has successfully developed independent study materials for the health sciences, business, and community services divisions. The team composition and work flow varied with each project. Some project teams worked more efficiently and effectively than others. The merits of using course development teams outweigh the disadvantages. The team approach to course development permits access to a wide range of knowledge and experience, fosters an exchange of ideas, cultivates collegial synergy, and develops the skills of its participants. □

References

England, E. The Design of Versatile Text Materials. *Open Learning*, 1987, *2*(2), 11-15.

Frame, J.D. *Managing Projects in Organizations: How to Make the Best Use of Time, Techniques, and People.* San Francisco, CA: Jossey-Bass, 1987.

Holmberg, B. *Status and Trends of Distance Education.* London: Kogan Page, 1981.

Kelly, M.E. Course Teams and Instructional Design in Australian Distance Education: A Reply to Shaw and Taylor. *Distance Education*, 1987, *8*(1), 106-120.

Kelly, M., and Haag, S. *Teaching at a Distance: Ideas for Instructors.* Waterloo, ON: Teaching Resources and Continuing Education, University of Waterloo, 1985.

Moisey, S.D., and Wright, C.R. *Roles, Responsibilities, and Workflow in a CML Development Project: A Look at Course Development Teams.* Paper presented at the International Conference on Computer Assisted Learning in Post-Secondary Education, Calgary, Alberta, 1987.

Perry, W. *Open University: A Personal Account by the First Vice-Chancellor.* Milton Keynes, England: The Open University Press, 1976.

Perry, W. *The Open University.* San Francisco, CA: Jossey-Bass, 1977.

Shaw, B., and Taylor, J. Instructional Design: Distance Education and Academic Tradition. *Distance Education*, 1984, *5*(2), 277-285.

Smith, C.L. Educators as Course Developers: The Key to Successful Microtechnology Integration. *Educational Technology*, 1987, *27*(8), 31-33.

Wright, C.R. *Course Developer's Manual.* Edmonton, Alberta: Instructional Development, Grant MacEwan Community College, 1987.

Wright, C., Zwicker, D., and Conroy, C. ShareTech: Nurturing Technological Change in Your Institution. *Educational Technology*, 1987, *27*(7), 38-40.

Figure 1

GMCC Work Flow Diagram

GMCC Work Flow for Instructional Development Projects

Program _____

Course Name and Number _____

Module Title _____

Course Developer _____

Word Processor Disk No. _____

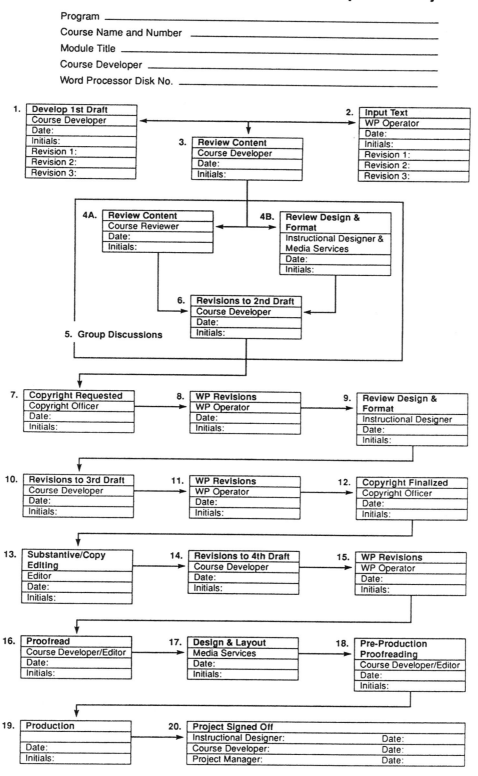

How Television Is Changing Thinking Patterns as We Move Toward Computer Controlled Global Communication Networks

Dennis M. Adams and Mary Fuchs

Computers and video technology are inexorably linked; you need one to view the other. By analyzing the patterns of the older technology (television) we may be able to anticipate future human-computer-video interactions. In fact, television could serve as a metaphor for future computer controlled technology-human communication.

We have now reached a stage where events outside of ourselves seem unreal if they do not appear on television. We are even coming to depend on it for perceiving the difference between what's real and unreal. American TV may be inconvenient, intrusive, and subtly harmful, but it is becoming a central condition of being over much of the globe. Video is now beamed 22,000 miles to a satellite in space and beamed down to any place on the planet at the speed of light. Unfortunately, imagination and intelligence are often missing from the programming it carries—and the language that TV uses is based on a very limited vocabulary, shattered sentences, and simplistic thinking levels. Although this may be helpful to those just trying to master simple language patterns, TV English is not very helpful for those who want to move beyond a fourth grade vocabulary.

Recent communications/computer alliances set forth a global strategy of using new technological combinations to bring visually based ideas and information to a world-wide audience. (The recent IBM, CBS, MCI alliance is just one example.) These multinational computer-communication combinations intend to serve up video as well as data. Unfortunately, the TV packages that are a main part of the mix (for the mass audience) do not point the way toward literacy, nor cultivate the kinds of language and thinking skills that children will need in a thought-intensive future environment. In fact, the miracles of our new technology may contribute to the erosion of literacy and language patterns on an unprecedented world-wide basis. Yet television has great holding power—and we all seem destined to spend more time watching the video tube than reading, writing, and speaking put together.

In the United States, more time is now spent on watching television than doing any other thing except sleeping. The majority of children and adults (in the U.S. and other industrialized countries) have come to depend on TV for intellectual perception as much as a minority has come to depend on junk food for nutritional sustenance. And, whether we like it or not, it's likely to remain that way. The problem is to maintain intellectual standards while using technology itself to turn around a technologically induced diminution of literacy and human visual perception. The difficulty is compounded when we are forced to handle the powerful synergism that will be progressively released as video merges with other technologies. As television itself undergoes radical change, we could find that it is a fine line indeed that separates the meaningless from the meaningful.

How Television Influences Children's Language Skills

Television points up the unsolved societal/educational problems of reconciling technological innovation, political democracy, economic monopolies, foreign competition, and a panorama of unmet educational needs. Heavy viewing seems to bring children evaluated as having excellent English language skills (writing, reading, and speaking) closer—in terms of test scores—to students with lower language skills. Our informal survey suggests that TV does this by lowering the scores of its brightest viewers and upping the scores of the lowest. American television typically uses a lowest common denominator formula to bring a mass audience a serialized story that focuses on the obvious—avoiding any linguistic or intellectual complexity. Hopefully, the refusal of many states to go along with watered-down, lowest-common-denominator textbooks may be extended by broadcast television.

In our survey of the language building potential of American television, we found that prime time programming was limited to about 4000 different words—with very little use of metaphor or other concepts requiring higher levels of thinking. For programming to aid in language development

Dennis M. Adams is Associate Professor, Educational Studies, and Mary Fuchs is Instructor, Mathematics Education, at the University of Northern Colorado, Greeley, Colorado.

would require at least a doubling of that number; the average high school or college class makes use of over 10,000 words. Lack of intellectual complexity and sentence length is another problem. Television scriptwriters tend to stick to obvious concepts and use sentences of about five words in length. Sentences in nonfiction books average over twenty—and even the worst book has to make some use of metaphorical thinking.

What students read and what they see and listen to on television affect their writing style and their thinking. When students watch television, to the exclusion of books, the result is fewer words and simpler sentences in their own writing. Since television programs contain primarily simple sentences or fragments, they have a negative effect on writing, language, and thinking skills.

Taking Time Without Giving Much in Return

Print and television are two media that compete for children's time and attention. But video is more seductive. Children frequently report finding television more life-like, easier, and desirable to spend time with than books. When similar groups have a chance to view or read about a particular subject, they find the work harder with a book. Not only do they comprehend more about the subject when reading, but they are forced to conceptualize at a higher level of abstraction.

The behavior of parents is a key element in the media choices that children make. This behavior strongly influences whether or not a child reads, and how much or what is watched on television. Parental TV watching lends credibility to the medium in general. And their programming choices assert a strong influence when it comes to specific programs.

Although we typically spend twelve years preparing American children to deal critically with reading, we generally neglect teaching critical TV viewing skills. And all of our graduates watch television—whereas perhaps only half read. Pulling the plug, in our opinion, is not practical, so we have to learn to make the best of it. We may not want to increase the number of televiewing homes, but we *can* direct viewing habits and reap some cognitive benefits from viewing. If we dismiss video as unworthy of our attention, then we may harm the child's ability to deal with a visually intensive future.

Some Television Has a Positive Social and Educational Effect

There are a few bright spots. For example, watching multicultural segments on "Sesame Street" can increase preschoolers' interest in playing with children from different ethnic backgrounds. Racial and ethnic differences on TV can be stereotyped by convenient stage mechanisms, but just as frequently television breaks down stereotypes.

To point out how ubiquitous American television has become and how it could contribute to global understanding we can look to *The Day After* (first broadcast in late 1983). The Russians who viewed the program (although it was not broadcast throughout the Soviet Union) reported being very moved by its content. This was one of the first programs where a large number of American parents and children watched a preselected program together—with educational advice from the National Education Association. Of course, intelligent TV viewing should not be limited to the few times that issues as powerful as nuclear war are dealt with. But this program did provide concrete evidence of what *can* happen when educators and parents work together on intellectually expanding television. By making the unthinkable thinkable—and abstract concepts concrete—*The Day After* is an example of the type of programming that can make a social and educational difference on a global level.

Interactive Television

Television (broadcast and cable) can perform a vital social-educational function in a world where ideas and information of the pictoral, print, and machine language variety zips between schools, homes, libraries, companies, and governmental agencies.

There are two new approaches that are now being successfully used to make television a more active tool for learning. The first uses videodiscs and videodisc players connected to microcomputers. Children use this video technology to enter the Louvre in Paris or take a moon walk with the astronauts. They can stop the action, explore an object of interest slowly from any angle, and even call up print information that describes the conditions under which the artist created the painting, or the moon rock was retrieved. The production and viewing of art and science in this way can help demystify the concept of creativity and bring us closer to our environment.

The second approach to combining television, language activities, and computing involves programs broadcast over the air. These programs, like PBS's *Voyage of the Mimi*, include booklets combining reading-writing activities and microcomputer diskettes that are sent to schools.

Broadcast television offers some intelligent programming that can be viewed by older children and adults. But the seven- to twelve-year-olds fall between the "Sesame Street/Electric Company" and the adult programming cracks ("Nickelodeon" and "Disney" are now on cable). It seems odd that even public broadcasting has no drama for children in this age group. The nature of American television is, however, influencing this age group throughout the world. The French news magazine *Le Point* recently carried a front page story for adolescent TV viewers (and others) entitled "Culture: How to Resist America."

American broadcasters need more of a chance to show imagination and intellectual depth for the entire spectrum of viewers around the world. Television writers, producers, and programmers should be aware of their world-wide impact. Hopefully, many of them will come to realize that we are all involved in selecting our future through the education of children.

When actual experience is not possible, television may be the best means for influencing the imagination, intellect, and the aesthetic values of children. Video is a powerful way to present information. And when it is coupled with informed interaction, either human or non human, it enables children to learn and grow.

As far as the educative potential of new media combinations are concerned, the jury has not even heard the evidence yet. Print, video, and computers may very well become complementary tools for learning. Computers run on electrical energy and the human brain on chemical energy. The human mind is not like a computer, but both transmit radio frequencies when they are functioning. And computers have the potential for controlling a type of hook-up between the brain's computer and other media. Does this mean that computers will learn to read our thoughts? That will take some time—at the moment the best computer can't even read our handwriting. Like any experiment, the educational results of computer controlled video-human combinations will be viewed down the road, rather than at the beginning of the effort. Of one thing we can be certain: rapid new communication tools will have more direct access to our thinking process, and they are beginning to be tested. Some of these tools will have a major impact on personal, national, and the international nervous systems. Various forms of computer-based video will increasingly command world-wide attention. Once we understand clearly the most effective way to learn a subject, we can teach a computer to do the teaching. Advances in artificial intelligence are resulting in microcomputer packages with decision-making abilities.

The computer-communications industries can set and accomplish the goals of contributing to the intellectual, aesthetic, and emotional maturity of viewers. To do this requires bringing together educators, media specialists, and scholars from many interrelated fields to produce intelligent programs that appeal to children. This is particularly important now that computer controlled technology is starting to give us the opportunity for interacting with television in various ways—taking even more time away from print media and actual experience.

In different combinations with computer-based technology, TV can take on new dimensions that will give video even more attention-getting power—and the potential for actively involving and expanding, rather than diminishing, human intellectual capacity. Although they have yet to be tapped, the technological opportunities for nurturing intelligent viewing are increasing at a surprising rate. This is especially true now that multinational companies (like IBM and AT&T) are in the process of designing new electronic networks where everything is linked to everything else on a global basis. The whole thought-producing process is poised for a quantum leap. It is time to consider how technology can be employed to eliminate illiteracy and ignorance—promoting human development on a world-wide basis.

"Programming" Is the Weak Link in Creating a Global Exchange of Ideas

If you think that we are in the midst of profound technological experiment and change now, just wait a few years. A revolution in global communications and information distribution is about to get underway. IBM's recent alliance with CBS and MCI (for a nationwide videotex service) and AT&T's move into computers recognize the need to merge the computer and communications industry. The result will be a global exchange of information and ideas that is beyond anything we can imagine today.

Language and patterns of thought, rather than computer skills, will be the key for dealing with the powerful synergism that results from the technological synthesis of software, computers, and communications. These new pipelines will carry moving images as well as data, print, and voice. The mix involves fiber optics, cable, satellites, infared transmission, new switchboards (to direct traffic), new software (to make messages understood) and, most importantly, new programming.

With international competition a major force behind technological development, it is important to come to terms with how people are manipulated. What exactly is the price, in human values, that we have to pay for our advances in science and technology? We have yet to come to grips with the potential social-educational implications of tapping these technological potentials. No one has spent much time looking at the effect of TV language and thinking on a mass audience—or the educative possibilities. It doesn't do us much good to have 155 channels of the same bland nonsense, no matter how much you can interact with it. And it certainly doesn't do other countries much good to suffer from this kind of cultural imperialism.

There is a gap between how people throughout the world view education (idealistically) and the way differentiated opportunities that are the reality exist for most of them. Television, at the moment, is the common denominator. Our negligence in using it effectively is not trivial—in fact it points to how we may misuse future technologies. American communications programming doesn't feel responsible to its world-wide mass audience, many of whom can't or won't read. There has to be a little inclination to use writers, directors, and producers who have both imagination and intellectual depth; otherwise, it will continue to diminish human culture and language skills.

A Large Dose of Very Human Intelligence Must Guide the Technology

Television is a perfect example of how technology, by itself, cannot cultivate or improve human learning, nor human nature. And it certainly doesn't carry with it a moral or ethical code that automatically points towards an improvement in the human condition. Divorcing the search for technological possibilities from human values and intellectual concerns results in devices that steal our time, and the time of our children, without giving much in return. Our short-sightedness allows an almost miraculous computer controlled technology—and the public airwaves—to be used (without much social conscience) for private profit. Far too often consideration of educative potential and public service is viewed (by insiders) as "public relations" image-building.

From the beginning this country has recognized the interconnection of technology, science, art, literature, and education with public funding. (All are now in the process of having their federal support cut.) President George Washington told Congress: "There is nothing which better deserves your patronage than the promotion of science and literature." President John Quincy Adams called for laws promoting "the cultivation and encouragement of the mechanic and the elegant arts, the advancement of literature, and the progress of the sciences." President John F. Kennedy quoted Presidents Lincoln and Roosevelt when he said that technology, education, humanities, and the fine arts were "far from being a distraction in the life of a nation. They are close to the center of a nation's purpose—and a test of the quality of a nation's civilization."

Television is a primary means of involving most of our nation in the humanities today. Bad writing, sloppy language, and limited vision may be rampant in American video communications, but nothing, it seems, can diminish the time we spend with broadcast TV. And there is no way to predict the global influence of such work. But in its present form it is not likely to compel much human language or aesthetic development—or much thought.

As this medium merges with other technologies, there is cause for concern. The problems that arise on the cutting edge of media technology involve not just the new communication-computer corporation alliances—but education, the arts, ethics, politics, psychology, medicine, biology, and philosophy. Like all machines (some of which date back nearly 10,000 years) the computer is a mechanical device for capturing the power in nature. The potential is there for enabling the future. At issue here is nothing less than preventing a technologically induced dark age of perception and literacy. But there is no need to limit ourselves to prevention. Whether or not computers ever achieve a level of true consciousness, they *can* become partners in teaching and learning. If we learn to manage today's technological miracles, we could go on to set the stage for extending the potential of the human mind (on a global scale) in ways never before thought possible. □

Computing and Telecommunications in Higher Education: A Personal View

Norman Coombs

Computer technology has drastically transformed my professional life both as a scholar and a teacher. I am a blind professor of history in the College of Liberal Arts at Rochester Institute of Technology, and in recent years I have come to rely on a PC with speech synthesizer to assist me. Besides using word processing for writing hand-outs, memos and professional articles, I connect with the university's main computer system using a modem and phone line. Students submit their work on electronic mail, and I even use a computer conference system to teach a class of off-campus students who can do all their work from home or office using telecommunications technology.

When I began college in 1950, I did my writing with a Braille slate and stylus. When I needed to produce something for class, I first read a line of Braille, moved my hands to the typewriter keyboard and typed what I had read. Jumping back and forth from Braille page to typewriter keyboard contributed to a large number of typos. When my typed document needed editing, I had to have a reader who could read back to me what I had previously written. I started graduate school in 1955 at the University of Wisconsin with one of the earliest tape recorders. I suppose it must have been a portable because it did have a handle, but it required an electrical power connection and weighed some 50 pounds.

Evaluation of Technology

When I began teaching in 1961, I was still limited to traditional and cumbersome ways of working. Student

Norman Coombs is Professor of History at the College of Liberal Arts, Rochester Institute of Technology, Rochester, New York.

work was turned in on paper, and I had readers reading papers and exams. Not only was this time-consuming, but it required our working according to a mutually agreed-on schedule.

When I first began using a PC and speech synthesizer in the early 1980s, I immediately found that it enabled me to write and edit work by myself. Often I preferred to have my secretary put out the finished version, so I learned to use a modem to transfer my document to the college computer and E-mail it to the secretary, saving the need for retyping.

I found electronic mail could be used to have students send papers to me via the computer, and have my speech synthesizer read them to me, giving me even more control over my own work. One of the first to send me material through electronic mail was a hearing-impaired woman. After receiving her grade by return E-mail, she wrote excitedly back saying that this was the first time she had ever "talked" with a professor without needing an interpreter. One of RIT's colleges is the National Technical Institute for the Deaf, and hearing-impaired students are in many regular classes. Normally, there is an interpreter seated beside the professor, signing for the deaf students. When these students want to talk with their professor in his office, they have to match their schedules with that of the professor and an available interpreter. Besides communicating with my hearing-impaired students via electronic mail, I encourage them to come to my office any time even without an interpreter. I turn on my speech synthesizer and put my PC into word processing mode, and we chat by passing the keyboard between us. We both come to value the immediacy of this direct communication.

RIT's library catalog has become computerized, and it can be accessed using a computer and modem from home or one's office. Being able to connect to this system with my PC and synthesizer was an exciting new privilege. I could never search a card

catalog by myself before. Of course, the first book I looked for was my own, and I was excited to "see" my book actually there in the catalog. The ability to use the same system to access other databases around the country has further enhanced my professional work.

Although preparing material for the class and grading student work is an integral aspect of teaching, it is not actually part of the instructional delivery. I have found two ways to utilize the computer more directly in content delivery. First, students are often not skilled at using a word processor or sending messages through electronic mail. I have a Zenith laptop which I can take into the classroom and display using an overhead projector. I now take one period to demonstrate to my classes what they need to know in order to write and mail their work to me using the college's VAX computer system.

Second, I now teach a history course for the College of Continuing Education for students off campus. They receive the course content by watching cablecast videos and reading the text. They engage in regular class discussions using telecommunications and a conference system, and they chat personally with me or each other using electronic mail.

Computer Conferencing

In 1985 RIT awarded me a productivity grant to explore the uses of computer conferencing and E-mail to add interactivity to a history telecourse. Using grant funds, we purchased VAX Notes as our computer conference system and purchased modems to loan students who wanted to access the system from their homes. The conference took the place of a classroom discussion. However, this was a discussion which was independent both of place and time. Not only could students do their work from home or office, saving commuting time, but the conference was available for access 24 hours a day.

Each week I posted a set of three to five topics on the current material, consisting of several questions. Students logged on to VAX Notes and attached their responses electronically to the relevant topic. I checked the postings for new entries several times each day and added comments of my own when appropriate.

Students and I both felt that there was more than the usual interaction between professor and student. It is unusual for a professor to single out a student in public and call him in for a discussion, but VAXmail made it easy to develop one-to-one conversations quite frequently. One student remarked that, as a result of being able to use this facility, this professor was the most helpful she had encountered at college. A questionnaire given to the pilot group asked them to rate professor helpfulness and availability compared to normal class settings. The students ranked that item 4.8 out of 5 and also scored electronic mail as the major factor in contributing to that process.

Students found that the computer conference did provide a genuine group discussion. "The computer aspect of the telecourse made it possible for me to compare myself and my progress to that of my fellow classmates," one member said. "I doubt that would have been possible otherwise." Another maintained that he would not "take another telecourse if a computer were not included." He found the phone and regular mail unsatisfactory for interaction. Others liked the freedom and independence the system provided for them to make their own schedule and set their own work pace. "I like listening to tapes," another commented, "watching videos and working on questions pretty much at my own pace." One of the few complaints came from a student who said, "I want to say that this course would be excellent for someone who has more self discipline than me. , ." Obviously freedom and independence require a degree of self discipline.

More Interaction

Most people are surprised to discover that computer conferencing encourages more personal and intimate sharing than does a face-to-face class setting. We all know that people often share personal things with relative strangers on buses and airplanes. Similarly, I found my students sharing in a way they seldom do in person. In a discussion of the impact of the Depression, several told touching personal stories about lingering habits of hoarding typical of their parents or grandparents. One young woman confessed to being on welfare and expressed her shame and hatred of it. A deaf student compared the impact on his self esteem of his losing his hearing with that of men losing their lifetime career jobs. Another man said he liked the conference system because he could express his opinions more openly. He held rather extreme political views and regretted that his friends had stopped discussing politics with him. He felt that on the computer people did not get as angry or personally involved. His wife, who also was in the class, said she liked the computer class because she did not have to read her husband's contributions.

The Notes conference also facilitated interaction with hearing-impaired students. The videos are captioned by the National Technical Institute for the Deaf. One hearing impaired woman said it was the first time she had really felt part of a class. She could readily understand what others were saying and could more easily contribute herself. In fact, she claimed it was her best academic experience in college. She had lost her hearing later in life and struggled to communicate with sign language. Others, who had learned through sign language throughout their schooling were less enthusiastic.

The computer conference class described here was given the same material and identical multiple choice exams as was another group of students who met in a traditional classroom. The computer section scored a higher mean grade than those in the traditional class. The control group scored a mean grade of 78.7 while the computer students averaged 82.0. This probably does not mean that computer conferences are better than class discussion. Rather, the use of the computer frightened away the below average student. The computer section also did have a grade point average slightly above that of the control class. This does demonstrate that telecommunications can be used effectively to bring a lively interaction into a telecourse and transcend the boundaries of time, space, and physical disability. □

Part III

Perspectives on
Educational Telecommunications

Technology: Implications for Long-Range Planning

David Foster

Introduction

Tomorrow's schools will have the ability to put computers with the power of yesterday's university mainframe on the desks of students for less money than today's micro. Other technologies (telecommunications, storage, etc.) will fuse with computer technology to make the student's desktop an incredible learning environment. Technology has an important role to play in the future of education. Most experts agree with that assessment, although not, perhaps, on the precise mechanisms of that role. Leaving educators and policymakers in a quandary—*what should be done today*?

Today technology is suffering from a few setbacks and faces an uphill battle. Henry Jay Becker (1987) concluded from his *Second National Survey of Instructional Uses of School Computers* that, "In terms of statistical patterns, computers so far have had only a limited impact on children's learning in school." The same has been said about telecommunications and other technologies. Education reports are quick to point out technology's failure to impact education, and, all too often, educators readily assume that technology has not lived up to its promise (National Governors' Association, 1986; Nathan, 1985; and Levin and Meister, 1985).

However, technology *has* lived up to its promises. It can do all the things it has promised, and it can do them very well. The problem is that technologies have outpaced our ability to use them, especially in education. Becker goes on to clarify his findings about the computer's lack of impact, noting that it "is due to the limited amount of computer equipment in the schools, the relatively brief experience provided to individual students, and the generally unsystematic use of software at the lower grades."

David Foster is a Senior Technology Associate at the Southwest Educational Development Laboratory, 211 East Seventh Street, Austin, Texas.

How should educators and policymakers respond to these findings? Alfred Bork (1987) put it this way, "Education worldwide has major and increasing problems. We have the capability, using the new interactive technology, to ameliorate these problems. But this will only happen if *we learn from the past to plan for the future*." [emphasis added] Technology will be a major force in solving education's problems if we develop proper long-range plans.

Technology has not been properly developed and implemented in the schools. Education's use of technology should be *needs-driven, function-oriented*, and *planned-for*. Too often it is not. As a result, educational technologies have left the forefront of attention in educational policy development—replaced by career ladders, teacher testing, kids-at-risk, etc. Implementing technology has been problematic for several reasons. It has been far more difficult, time consuming, and costly than expected for educators to take full advantage of technologies' capabilities for education. Many specific technology projects have been oversold and improperly implemented. There has been [and is] ongoing resistance to, and fear of, change. There has been a lack of training for teachers and administrators. Costs have been prohibitive for many states and districts. And, most of all, states and school districts have suffered from a lack of vision and planning.

We have, meanwhile, learned a lot about development and implementation. We are getting a handle on costs—at least we understand that development won't be cheap or easy. These facts should not be surprising to anyone. We were told. Back in 1984, David Moursund told us that "while CAL [computer assisted learning] has tremendous potential, the cost of achieving that potential is high and the timeline is long. The current and next five-year impact of CAL upon our total education system will be modest." And so it is.

But, the future for technology is improving because of advances in design and production. Such advances have caused technologies to come down in price while dramatically increasing in capabilities; this trend will continue well into the 21st Century, making technology even more attractive to education. Nevertheless, a widening gap exists between demonstrated effects and existing potential.

We cannot let past problems with implementation dictate a limited role for technology in the near future of education. We should use the past to inform our plans for the future. Despite its recent let-down, technology's role in education has not been abandoned. Technology will touch every aspect of education and change the system as we know it. Why? First, need; second, cost-benefit ratio. Already, technology *is* being used to solve teacher shortages, provide cost-saving curricula, carry out administrative functions, save on paperwork, improve students' low achievement test scores, involve gifted and talented students in advanced learning activities, improve information flow and processing, and bring at-risk kids back into the education process. State-level decision-makers should first examine the overall education picture as it relates to technology. They must see where their states are; envision where they want to be; and begin to draft plans for using technology to get there.

Technology and the Overall Education Picture

How is the word *technology* being used here? There are several problems in developing a specific definition for the word technology. We often use the word to explain those things we do by using machines that we don't understand. When our new car senses our presence at night and turns on its interior lights we say *with pride*, "That's technology!" When said car's electronic ignition doesn't work the next morning we say *with pride forgotten*, "That's technology!" Technology has, thus, become an emotion-laden term, and we often find educators primed by the first experience in heated disagreement with educators disillusioned by the second. Clearly, defining technology is not the simple task one might expect, yet for policymakers the definition is critical and must be context specific. For the term technology is used—almost always in an emotionally charged context—to denote products or processes that we do not quite understand.

A specific operational definition of technology is not developed here. However, we know technology is an *evolving* process that *enables* the development of many products and procedures that will exert an *inevitable influence* on our concept of education. It is also being rapidly suffused into every aspect of the world-of-work. Its use will be a *basic skill* for most, if not all, of tomorrow's good jobs.

Technology Is Evolving

Technology is not static. As we will discuss later, the stand-alone microcomputer that is increasingly prevalent today in classrooms bears little resemblance to integrated technology-based learning systems that are already becoming available.

The new focus on technology in 1987 emphasizes solutions to educational woes that have continued to plague America as the wave of reform passes. This focus is on the fusing of many existing technologies. Deloitte Haskins-Sells has noted that, "it has become apparent that knowledge-based systems, which integrate logic, audiovisual, voice and data systems, will play an increasingly important educational and economic role in K-12 education."

Out of existing capabilities will come the technologies of tomorrow. The boundaries between technologies are breaking down as integration and networking between systems develop? And as those technological boundaries gradually break down, so too will other socioeconomic, geopolitical, and informational boundaries. Indeed, we can look to an integration (or at least an aggregation) of resources and ideas from many sectors. We will see telecommunications systems and computer systems exchanging input while providing a variety of integrated services to students, teachers, and administrators. Take, for example, just one aspect of the system—the computer—and look ahead to get a glimmer of technology's untapped potential.

The instructional computer of tomorrow will play an important role in education. One only has to look at existing hardware and software capabilities, and at available peripherals and combine them to get a view of tomorrow's instructional computer system—and the view doesn't resemble today's micro. The instructional computer of tomorrow will:

KNOW
1. Various instructional strategies and use them prescriptively.
2. Learning theory and learner diagnostics.
3. How to make inferences using inductive and deductive reasoning.
4. Content areas better than any single living expert.

SENSE AND RECOGNIZE
1. Visual images.
2. Sounds and voice.
3. Physiological state of its operator.
4. Input from many devices, including the human voice.

LEARN
1. The learner's relevant prior knowledge.
2. The history of the user, including likes, dislikes, hobbies, friends, and pets.
3. The student's instructional level and learning orientation.
4. New information. (These systems will be able to collect information and data about the individual learner, process this information, and respond according to an ongoing analysis of this information.)

PROVIDE
1. Interactive, full-motion video and multi-lingual audio (hypermedia).
2. Speech recognition and verbal response.
3. Elaborate branching that allows the learning sequence to be governed by either student or curriculum objectives (hypertext).
4. Data at the individual, classroom, school, district, and state level (at the speed of light until administrators develop a way of slowing it down).
5. Decision-making support for all levels of education.

USE THE ABOVE CAPABILITIES TO PROVIDE CURRICULA THAT

1. Provide in-depth prescriptive diagnostics to the teacher using proven interventions.
2. Use embedded testing to determine student mastery and instructional level.
3. Conduct individual learning research on users.
4. Provide data and record keeping automatically, in appropriate form, for teachers, parents, students, schools, districts, states, and national assessments.
5. Communicate automatically with worldwide databases and sources of information—updating both its information base and its software capabilities.

The point, simply, is that the capabilities of emerging and new technology-based systems are not much like existing, classroom-tested products. These new capabilities will redefine the ways teachers and students act and interact in the classroom. This new generation of instructional tools already is redefining how we use evaluation data. We cannot, for instance, afford to say whether the new or emerging products work or do not work based on the evaluations of other, earlier generations of products. Instead, we must begin to ask different questions about the new capabilities and the new relationships between classrooms and technology-based tools. Too many existing evaluations come from pilot studies and are seriously confounded by *implementation* problems. Evaluations of a satellite-delivered course that experienced initial audio problems can not be generalized to the same course once these initial problems have been fixed.

Technology Enables

Technology is creating what Christopher Dede (1987) calls "empowering environments" referring to its ability to function as a cognitive enhancer. Computers are able to handle and manipulate information and knowledge in increasingly sophisti-

cated ways. Terms like "hypermedia," "microworlds," and "artificial intelligence" are being used to describe the capabilities of current-generation computers to handle a variety of information through structured symbolic manipulation—increasing our mental productivity is the result. The products and procedures created from or based on these and other evolving technologies are many. They have many different faces and applications that can address the question asked by concerned educators, "What is technology providing the education system?"

The results of technology can be: a cost-effective instructional delivery system, an intelligent tutor, a classroom management system, a statewide communications system, a major sector of a state's desirable job market, a content area deserving of curricula, a microworld that simulates reality, and these results are increasingly a part of the everyday life of all Americans. They will affect the way educators collect data, manage information, write job descriptions, and design school buildings. They will use, validate, and change learning theory. And they will play a major role in the employability of the graduates from our schools.

Technology Exerts an Inevitable Influence

Pro-technology critics of our education system have not helped technology's image with their threats. Legislators, however, are beginning to look to technology to make our education system more competitive with foreign systems. The following viewpoints are worth considering. Lewis Perelman (1986) believes that,

> The age of schooling is over. A new, post-industrial 'learning enterprise' is about to replace the outworn infrastructure of industrial-age education. The technology we call 'school' will have as much place in the 21st century's learning system as the horse and buggy have in today's transportation system.

> The economic, social, and technological trends pushing critical choices about learning to the top of the public-policy agenda include:
> ○ The quality, value, and relevance of most educational services are declining, but the development of a post-industrial economy means that the population has growing, unmet needs for learning.
> ○ The costs of education and training have become the biggest single item of spending in the U.S. economy and are rising several times faster than personal income.

○ The demand for government subsidies to education is exploding, but the deficit crisis is going to require fiscal austerity.

○ Modern information technology can be used to increase the productivity of learning, but an entrenched educational establishment obstructs its application. (p. 13)

Perleman's rhetoric might seem a bit strong, but let's look at what key policymakers are saying. *Education Computer News* (June 17, 1987) quotes one Washington education observer who states, "People in the legislative and executive branches are finally waking up and saying, 'Hey, the system we have is worthless, and it has to be fixed.' " He goes on to note, "The competitiveness issue has helped, in that it forced us to recognize that our rivals overseas are kicking us all over the map where education is concerned."

One "awakened" legislator is Senator Edward Kennedy, who describes our education system this way, "The economic battles of tomorrow are being fought in the classrooms of today and the news from the front is not good. In survey after survey, American students are at the back of the class in math, science and foreign language achievement." The Senator is one of several federal legislators introducing technology-based education legislation. Both the House and Senate versions of the Trade Bill (H.R. 3 and S. 406) have technology-based education initiatives, legislation designed to provide "Education for a Competitive America." Many states also are considering the role technology should play in their long-range legislative and regulatory agenda.

Long-Range Planning

Without debating the condition of our education system, we can determine that technology will play an important role in the future of education. It is up to business and education to work together to plan, develop, and implement successful technology solutions to nagging educational problems. And it is up to state governments to legislate the necessary contexts and incentives to insure timely and successful applications. Planning for this eventuality must begin by determining where we are now, with an eye to the future. Not just what *technologies* exist, but what elements of the system will or could be affected by technology.

Each state must:

1. Determine its current **situation** with regard for the future. What technologies are out there already? What do we know about their effectiveness? Will new advances offer significant differences? How is the current education system struc-

tured? Does that structure facilitate change?

2. Determine its educational **problems**. Are students competitive nationally? Internationally? Do the people in your state have increased expectations for education? Are these expectations being met? Teacher shortages? Effective training? Funding? Equity issues? Drop-outs? Etc.

3. Determine what **needs to be done**. What will put you on a successful path to the future? How does technology fit in?

4. Determine the reality-based **options/solutions**. What are the cost-benefit ratios for various technology-based solutions to your rural, equity, quality, training, administrative, information, and advanced-course-offering problems?

5. Develop a **plan**. Planning should be based on analyzed options for solving identified problems and should include the remaining points.

6. Provide for an **implementation** process and necessary **funding**. Plans must be implemented before they can do any good. The plan should provide the implementation procedures (training and funding are necessary components).

7. Provide for an **evaluation** that guides and informs the process. There should be formative and summative evaluations that are tied to educational needs and goals.

8. Build in **modification** capability. Any long-range plan involving technology should be designed to take advantage of evolving cost and performance benefits and to correct itself where necessary.

9. **Share** information and, when possible, resources, development, and activities. States cannot afford to learn the same lessons over and over again.

10. **Think of the future as a moving target**. Planning can not be a static process. We are always moving into the future.

Decisionmakers' Previous Recommendations

Many task forces have tried to address the issue of "the role of technology in education." The National Task Force on Educational Technology (1986) reported to the U.S. Secretary of Education that:

First, today's schools need to gear up to improve traditional delivery to the maximum that technology-based education helps make possible. They need to prudently acquire hardware and software and apply them effectively. One work station for every ten students is an achievable goal and an important step in the right direction.

Second, in improving current learning practices through technology-based education, the schools will need to pay careful attention to planning, financing, teacher education, and improving the curriculum and instructional practices. (p. 12)

The National Task Force went on to recommend that education needs to be transformed into a "new environment" designed to fit changing requirements. And they state that "The intelligent use of technology can be of enormous help as we strive towards these goals."

The National Governors' Association Center for Policy Research and Analysis made the following recommendations for states in their report, *Time For Results: The Governors' 1991 Report on Education*:

1. Encourage all school districts to develop written plans regarding the use of various technologies, prior to purchase of any equipment.
2. Encourage all prospective teachers to learn about effective and emerging uses of technology in their respective curriculum areas.
3. Encourage every college of education to develop relationships with as many schools that use technology as possible.
4. Help school districts and schools write informed, appropriate plans on the use of technology.
5. Help school districts, schools, and universities develop and establish continuous training programs on the appropriate uses of technology and ways to incorporate it into their curriculum.
6. Help all states share data on costs and achievement from experiments being conducted within their respective states.
7. Aggregate purchases and establish wider markets.
8. Provide financial incentives and assistance for district planning for the use of technology by providing support for the purchase of equipment.
9. Provide for establishing and improving training programs for educators about cost-effective ways to use educational technology.
10. Assist school districts that are willing to experiment with ways to restructure the school environments to increase educator productivity by using various forms of educational technology.
11. Encourage greater cooperation among the states through the creation of consortiums and other technical assistance arrangements.
12. Establish independent institutes for research and demonstration of technology in education, modeled on the National Science Foundation (NSF).
13. Recognize each state's most creative technology-using educators.
14. Make technology more available for students from low-income families.

States Begin Long-Range Planning

Making lists of generic recommendations is one thing; improving or changing the system is another. Individual states seeking to identify the role technology will play in their long-range education planning are faced with a series of complex issues. Some states are facing the issues as they begin their planning efforts.

First, a state must determine some guiding principles. New York (1986), for example, developed three major goals for their long-range technology planning: excellence, equity, and efficiency.

Second, a state must determine the specific areas to be assessed. In Texas, for instance, state planners are looking at how technology may change or affect:

○ Instruction and instructional delivery
○ Textbook definitions (Should software be included?)
○ Inventory procedures
○ Textbook depositories
○ Materials acquisition
○ Teacher certification (Should teachers be required to have training in teaching via telecommunications? Should they be able to provide lessons by satellite?)
○ Personnel requirements and qualifications (Should non-teachers be used in classrooms receiving instruction via satellite or fiber optics?)
○ Personnel boundaries (How will new roles affect traditional "turf?")
○ Existing regulations concerning business-school relationships (What happens when a school leases an ITFS system to a cable TV company, shares fiber-optics with a phone company, or develops software with IBM?)
○ Role of intermediate agencies (Do service centers need uplinks? Will much of what they do be done by telecommunications?)

○ Accreditation process (Will information for accreditation be provided and processed by electronic networks?)
○ Technology's interaction with other trends (Will it herald a move from process-based to achievement-based accreditation?)
○ The need for additional education indicators (Will technology create a need for new indicators beyond existing standardized tests?)
○ Implementation issues (How do we turn a long-range plan into something more than a plan?)
○ Evaluation issues (What type and role should be developed for future evaluations—summative or formative? We know technology works—now what works best? In what situations? Why? What are the implications for future developments, learning theory, policy, etc.?)
○ Finance (What are the cost-benefit issues?)
○ Training issues (What training will need to be provided on and by technology?)
○ Management
○ Facilities

In other words, Texas is asking: "How does technology fit into every aspect of the overall education picture?"

The Texas Education Agency has funded Coopers & Lybrand to develop a long-range technology plan for a state public school telecommunications delivery system. Coopers & Lybrand's planning strategy could serve as a model for other states to follow in the development of any long-range technology planning. They plan to:

○ Thoroughly investigate the current and future information delivery needs of the Agency, the state's Education Service Centers, and the local school districts through both surveys and on-site visits.
○ Research present and potential technologies for the delivery of information.
○ Study the literature on the possible uses of technology for public education.
○ Develop several alternative configurations of technology-based systems that will be capable of delivering the range of information services needed to support instruction and management of the public school system.
○ Define a set of cost ranges for each alternative.
○ Develop a series of projects with estimated timelines that will be designed to establish the proposed system and services.
○ Aggressively seek opportunities that may

lower costs of current systems and improve service.
○ Focus on functionality, flexibility, and existing resources.
○ Develop the most cost-effective approach that meets quality standards.

Such planning requires experience and expertise in both technology and education as well as awareness of political and financial realities. For further information on Texas's planning efforts contact either: Mr. Hans Wagner, Coopers & Lybrand, 1800 One America Center, Austin, TX 78701, (512) 477-1380; Dr. Geoffrey Fletcher, Director, Educational Technology, Texas Education Agency, 1701 N. Congress Avenue, Austin, TX 78701, (512) 463-9807.

Technology Will Force Fundamental Changes

Sir Clive Sinclair (1985) has observed that, "It often seems that each new step in technology brings misery rather than contentment, but this is because it brings change faster than benefits—and change, though often stimulating, is always disturbing."

Technology cannot be applied successfully to education's most intractable problems without "disturbing" the system. Implementing technology in the schools will change personnel patterns, union relations, certification, and even our definitions of curricula. Planning, policy, and implementation procedures must be conducted carefully, for they will affect acceptance and thus control outcomes. Unions, for example, may look at satellite-delivered courses either as a way of guaranteeing a certified teacher in the classroom or as a way of replacing the teacher in the classroom. How we proceed will affect attitudes and, ultimately, success.

Fundamental changes in all aspects of society are resulting from technology, and education will not escape. The long-term effects will be several.

1. Technology will exact more front-end costs for development, and it will take longer to reach its payoff. *Possible Result*: The need to plan for both the "birth" and "death" of technology. While some states are still planning to get their initial computers in the schools, others (such as Arkansas) are concerned with getting rid of existing computers to make way for new, more powerful ones.

2. The cost of developing curricula may be well over one million dollars per course, but the per pupil cost will be relatively low. *Possible Result*: The need for collaboration across state lines, both in development and purchase. Entities like

Arkansas' *Project Impac* and New York's *BOCES* will proliferate and develop collaborative relationships.

3. Research on technology will change its focus. *Possible Result*: Planning, implementation, training, funding, and numerous other utilization issues will need to be dealt with. For example: There is plenty of research proving that television can be an effective instructional tool, but television isn't having much impact in classrooms. *Possible Result*: Research will need to move on to such questions as: What works best? How? With what? In which way? We don't need another simple control-group comparison of "learning gains."

4. U.S. education will quit using secondhand "technologies" if it wishes to remain competitive. *Possible Result*: Business and industry working closer and in new ways with education.

5. The system will change. Education has maintained its "factory model" homeostasis for more than 50 years. Technology has no dependence on this model. *Possible Result*: A need for schools to keep pace with our changing educational needs for their own survival in a competitive market that offers new alternatives to traditional schooling.

6. Our knowledge-based information society is creating a need for life-long learning and re-training. *Possible Result*: A need for schools to expand their role to balance community needs and economic realities. Adults may be a necessary client.

As our educational environment changes, telecommunications and computer technologies will be there. The instructional computer will function as a "learning doctor"—one with humor, patience, and understanding. It will be a computer that works very hard and very well at being an effective and entertaining educator and teacher's aide. More important, it will free the teacher to give individualized help, thereby providing the very important human side to an increasingly productive, but technology-based, educational system. It will provide Intelligent Computer Assisted Instruction (ICAI), simulations (or microworlds), challenging programming languages like LOGO, productivity tools (word processors, spelling and grammar checkers), and expert knowledge via text, sound, graphics, and full-motion video.

Telecommunications will also play an important role. Satellites will bring top scientists and educators into the classroom live. Fiber optics and microwave will allow rural schools to share teachers while maintaining interactivity and classroom size. Technologies will be integrated to provide communications, information, data processing, training, instruction, and many support services.

Such capabilities have all been demonstrated and exist today in research laboratories, universities, and industry. American education deserves more than second-hand technologies and a competitive American education system will only result from the appropriate use of state-of-the-art tools.

Conclusion

What guiding principles should shape the planning process? We have learned from the past that decisions concerning the implementation of technology should be needs-driven, not "bells and whistles"-driven. We must look beyond current needs to the future; we have to understand the capabilities of technology in terms of both existing and future needs.

Any long-range plan should not underrate current or future technologies—they are *very* usable. Existing technology has been under-used and has still proven itself. We already can provide instruction via computer, coaxial cable, fiber optics, microwave, satellite, and various other technologies—and we can do it very well. Educational technology will provide live interactive instruction to remote locations via microwave, satellite, coaxial cable, and fiber optics. It is the most likely option to provide educational equity to our small and rural schools, solve teacher shortages, and improve our educational standing internationally. For, unless world philosophies change, our education system must maintain or improve our competitiveness in the world economy. We must plan how we will use technology to improve our schooling process in the future. And the future is now.

Summary of Suggestions

We must plan carefully. Technology will continue to shape our processes and systems of schooling. Whether our investments and energies are well spent or misspent depends on how well we anticipate and plan for the natural consequences of change. Chris Dede (1987, in press) has projected that various technologies will "facilitate the participation of every person associated with the educational process—learner, teacher, administrator, employer or parent—in shaping instructional outcomes." We need hardly add that such individuals can extend their participants among states, as well.

State-level education decisionmakers may choose to work together to create their own "empowering environments." If they choose to extend their participation across state boundaries to shape instructional outcomes, states should:

1. Take a systems-based approach to planning. All aspects of the education system will influence and be influenced by technology.
2. Learn from technology's past in education. Resist using this history to justify, rationalize, or maintain resistance to change.
3. Develop a vision of what education can and needs to be along with a vision of what technology might contribute to education.
4. Determine where education is now, how it got there, and the role technology has [is] playing.
5. Cooperate with business, industry, and other states to:
   ○ share research- and practice-oriented information
   ○ coordinate ongoing and new research efforts
   ○ aggregate buying of hardware/software, other equipment, or supplies
   ○ cooperate in developing projects, programs, and applications
   ○ develop multistate certification and accreditation procedures for courses broadcast across state boundaries.   □

Resources

Becker, H. Using Computers for Instruction. *Byte*, 1987, *12*(2), 149-162.

Bork, A. The Potential for Interactive Technology. *Byte*, 1987, *12*(2), 201-206.

Center for Learning Technologies, New York State Education Dept., Albany. *Learning Technologies and Telecommunications in New York State*. ERIC Document No. ED275297, 1986.

Dede, C. Empowering Environments, Hypermedia, and Microworlds, in press.

Levin, H.M., and Meister, G.R. *Educational Technology and Computers: Promises, Promises, Always Promises*. Stanford, CA: Institute for Research on Educational Finance and Governance, 1985.

Moursund, D. *The Computer Coordinator*. Eugene, OR: ICCE, 1985.

National Task Force on Educational Technology. *Transforming American Education: Reducing the Risk to the Nation. A Report to the Secretary of Education*, 1986.

Neill, S.B. *High Tech for Schools: Problems and Solutions*. Arlington, VA: American Association of School Administrators, 1984.

Perelman, L. Learning Our Lesson: Why School Is Out. *The Futurist*, March-April 1986, 13-14.

Task Force on Technology. *Time for Results: The Governors' 1991 Report on Education: Supporting Works*. National Governors' Association Center for Policy Research and Analysis, 1986.

This article is based on work sponsored by the Office of Educational Research and Improvement, U.S. Department of Education, under Contract 400-86-0008. The content of this article does not necessarily reflect the views of OERI, the Department, or any other agency of the U.S. Government.

Why Information Technologies Fail

Diane M. Gayeski
Contributing Editor

Introduction

Every decade has seen the introduction of at least one new communication technology which was predicted to radically transform informational and instructional communication: In the 50's it was film, in the 60's it was broadcast educational television, audiovisual aids such as filmstrips, slides, and overhead transparencies, and the new technology of programmed instruction; in the 70's, it was videocassettes, remote-access audio and video, and computer-assisted instruction. Today in the 80's we've been introduced to videotex, interactive video, teleconferencing, electronic mail, and recently, artificially intelligent teaching systems and job aids. **Has instructional technology succeeded?** Some think not. "With few exceptions, instructional technology has failed to live up to expectations" (Van Wyck, 1976, p. 291). Looking back over this history, we can point to technologies that have clearly succeeded, others that have perhaps been superceded, and yet others that have frankly failed. Is there a pattern to the successes and failures that may be helpful in predicting the successful adoption of technology and in designing more appropriate instructional vehicles?

Some Failures

Educational Television

Perhaps the most visible and memorable failure among instructional media is broadcast "educational" TV (Berkman, 1976; Gordon, 1976; Shorenstein, 1978). In the early 70's a number of government and private grants supported the production and distribution of television programs meant to replace classroom teaching. Such commercial ventures as "Sunrise Semester" did manage to capture small audiences who were motivated enough to watch these early-morning lectures, and

Diane M. Gayeski is Associate Professor and Chair of the Graduate Program in Communications at Ithaca College and a Partner in OmniCom Associates, a communication analysis, design, and production firm based in Ithaca, New York.

a number of universities and community colleges followed this model (Stern, 1987). What we now know as public television was first envisioned to be educational TV or "ETV"; the concept was to find superior teachers and broadcast their lectures to a wide geographic area so that students might be exposed to the best instructors available.

When carefully designed and produced, these broadcast lectures were effective: for instance in the Washington County, MD experiments funded by the Ford Foundation, students in the classes taught through the cable TV system showed remarkable gains in achievement (Saettler, 1968). The British Open University provided a successful model that many U.S. institutions followed in attempting to offer courses or even entire degree programs over the air, but many, such as the University of Mid-America, failed (McNeil and Wall, 1983). Cable systems, microwave hook-ups, and ITFS (instructional television fixed service) have all been explored by a large number of school districts. Although today there seems to be a resurgence of interest in this form of teaching, with a number of universities and school systems offering telecourses (Stern, 1987), it will be interesting to see if this technology takes hold the second time around.

Teaching Machines

Based on aspects of behaviorist psychology, educators embraced programmed instruction enthusiastically during the 1960's. This technique of individualized instruction was mediated through self-paced workbooks, branching or "scrambled" books, and in an audiovisual format through a host of teaching machines. Although this technology, too, proved itself effective and efficient through both research and field trials, one would be hard-pressed to find a "teaching machine" today (Kearsley, 1984). While it might be argued that they have been replaced by more sophisticated devices such as computer-based training and interactive video, the truth is that they disappeared at least a decade before these newer technologies were widely available.

Dial Access

Systems which allowed individual access to audio and/or videotapes at remote stations through a wired system were popular on campuses and schools in the early 1970's (Hempstead, 1976; Nixon, 1970; Schacter, 1975; Singer, 1970). The concept was that students could dial-up lessons from dorms or from carrels in the library. Another related rationale was that instead of trying to find room for large classes to meet on campus and paying for a number of instructors to teach repeated sections of basic courses, universities could enable students to watch videotapes of the lecture individually. Additionally, instructors could call up audiovisual material within their classrooms from a central distribution center. Even before this technology was made impractical by the advent of small, easy-to-use audio and videotape players, it was not used heavily or well. Although articles on this technology were common in the early 70's, not one could be found that has been published in the last ten years. Today, many of the expensively wired systems are still in place and could be functional, but the carrels and classroom controls have been removed due to lack of use.

Videotext

A newer arrival on the technology scene is videotext. These systems, introduced in the early 80's, enabled text and graphics to be transmitted either as part of an unused portion of a television signal (teletext) or over phone lines (videotex). Although these systems were designed to facilitate commercial transactions (such as buying airline tickets or engaging in home banking) there were informational and educational applications available as well. System developers, such as AT&T, envisioned on-line drill and practice lessons and encyclopedias which could be searched by keyword. Despite large sums of money invested in field trials, these systems have failed (at least in the U.S.). Salerno (1985, p. 132) described this group of technologies as "products in ceaseless search for a market" and remarked that the "problems of fitting it into the real world are prodigious." Bolton (1983) reported that as users gained more experience with a videotex system, Channel 2000, the lower their rating of the technology became. Although many on-line services are functioning successfully as bibliographic search tools and bulletin boards for special interest groups, the educational component envisioned has never been adopted in the United States.

Interactive Cable

Innovations that made possible two-way cable television promised a means to enhance and possibly revive broadcast educational television. Perhaps the most noteworthy is QUBE, the two-way cable system developed in Columbus, Ohio. Like earlier forms of broadcast instruction, pilot projects found that students learned as well or better from interactive cable programs as through traditional classroom teaching (Robinson and West, 1986). However, the educational as well as civic and recreational programming of QUBE never was accepted widely by audiences. After several years of struggling, the QUBE system and other two-way cable operations suspended interactive programming.

Some Successes

Video

Lest we become too discouraged by the preceding discussion, we should recognize the communication technologies that *have* been widely adopted in education and training. Perhaps video (not "television") is one of the most outstanding successes. Since about 1975 when the standardized, easy-to-use 3/4 inch U-Matic format became popular, video has been used extensively in education and training. The technology allowed not only easy and cheaper playback of generic programs, but facilitated local production. Today, non-broadcast video is a four billion dollar a year field (Brush and Brush, 1986) and video is the most widely used medium (even more popular than lectures) for corporate training (Lee, 1987).

Visual Aids (Business Slides, Overheads)

Although perhaps not as glamorous as some other technologies, projected still images in the form of slides and overhead transparencies are widely used in instruction and information. In fact, in total dollars spent, slides remain the most popular information medium: Organizations spent about $4.4 billion for production and equipment in 1986 (Hope Reports, 1986). As new means for computer generation of graphs, charts, and artwork become less expensive and easier to use, the traditional forms of visual aids such as slides, overheads, and handouts are becoming even *more* widely applied.

Questionable

Teleconferencing

Teleconferencing (video-based communication via phone lines or satellite broadcast) has been available for some time, but falling equipment costs and relatively inexpensive satellite time have made this medium more accessible to a wider range of organizations. However, video conferencing is "not materializing as fast as some have predicted and others have wished" (Hope Reports, 1986). Although there were predictions that teleconferencing would replace a great deal of business travel and facilitate home-based work, these have not been fulfilled. Substituting face-to-face meetings with teleconferences would dramatically shift the managerial and social patterns of organizations (Albertson, 1977). People *like* to travel, for the most part, and need personal contact with colleagues. Teleconferencing, however, *has* enabled wider participation in discussions and decision-making by making it feasible for larger numbers of middle managers and line workers to interact with managers and co-workers at a distance. Additionally, many professional organizations are using the technology to communicate special sessions or parts of their annual conference proceedings to members across the world.

CAI

Although computer-assisted instruction (CAI) was available in the 1960's, its more widespread adoption in education and training did not occur until the advent of the microcomputer. Today more than half of all organizations with 50 or more employees use CBT (Lee, 1987), and microcomputers for occasional tutorial use are common in elementary and high schools. However, according to most researchers in the field, the power of this technology has yet to be exploited: most programs are rather mundane tutorials on using computer software or "electronic flashcards" for educational use (Merrill, 1985). Merrill describes the stereotypic interactive instruction paradigm: "(1) Present a page of text (which may include graphics) to the student; (2) Ask a question; (3) Provide feedback on the correctness of the student's answer; if the answer is incorrect, provide remedial material (which is sometimes omitted); and (4) Repeat this cycle" (p. 20). Merrill reflects the disappointment of many researchers in new instructional technologies, and contends that some of these strategies are hold-overs from 1960's programmed instruction, which have been found to be boring.

Few schools and colleges use CAI on a systematic basis. Yet, some standardization into at least three major microcomputer formats (IBM, PC, Apple IIe, and Macintosh) has made generic software relatively inexpensive and available, and easy-to-use authoring tools are facilitating local production of custom CAI by non-programmers.

Interactive Video

This technology, combining the powerful attributes of sound, motion, color, audio, and tailored information via branching presentations, has been perhaps the most widely touted medium of this decade. However, interactive video for education and training isn't expanding as quickly as was initially predicted (Eikenberg, 1987; McLean, 1985). In fact, interactive video use essentially did not expand from 1986 to 1987, with 14.7% of U.S. organizations using interactive video to deliver training (Lee, 1987). Many of these organizations have produced or bought one program to test the technology, but few have adopted it as a widespread in-

formation vehicle. The perceived cost and complexity of the medium, added to the lack of standardization, has made many potential adopters wary. Organizations are just now getting interested in learning to use their own resources to develop custom interactive programming (Gayeski and Williams, 1987).

CD-ROM

The compact disc that has captivated consumers of audio recordings has emerged as a medium which can store not only audio, but computer programs, graphics, animation, and video as well. Although the CD-ROM technology is a few years old already, it has yet to see wide adoption. As in many other technologies, the "Catch-22 dilemma of hardware and software is hampering growth" (Brewer, 1988, p. 26). People won't buy the hardware because there isn't much CD-ROM software; on the other hand, developers are reluctant to produce media for which a large number of delivery systems are not available. In addition, the competing and mutually incompatible standards of CD-ROM, CD-V, CD-I, and DVI cloud the industry. In its favor, CD-ROM technologies offer huge storage capacities, multi-media display of information, relatively inexpensive mass production of software, and compact hardware which can be built right into a PC substituting for a conventional floppy drive. The question remains: how is this better than a stack of floppy diskettes, a few phonograph records, and videotape or videodisc?

AI

Although not a media technology *per se*, artificial intelligence techniques have a number of significant applications in education and training. Natural language front-ends mean that users are not restricted to communicating with a computer in specific codes, but rather, can use ordinary phrases to input information. Expert systems can serve as intelligent job aids, capturing the expertise of leading researchers and practitioners and making their diagnoses and recommendations available through a computer system. AI techniques promise to offer truly interactive tutorials which can develop a model of the student and select strategies based on that developing relationship (Kearsley, in Gayeski and Williams, 1985).

Although these techniques have been demonstrated successfully for over a decade, they are still rare in actual practice. Salerno (1985, p. 134) comments, "First, behind the true expert system there really is an expert. The knowledge engineers who develop these systems must work closely with the expert to formulate the rules that make up the AI program. Few experts are eager to devote themselves to making a machine smarter, and the knowledge engineers are in equally short supply. Add to this the need of AI programs for a special language and an expensive machine, and the slow growth of this promising field is not surprising."

Possible Reasons for Failure

Technophobia

Many enthusiasts of new information technologies label their more cautious colleagues as technophobic or conservative. "The reasons behind the perceived stall [in interactive video] are blamed on numerous things, but frequently cited are conservative attitudes towards an unknown entity" (Eikenberg, 1987). However, generally, non-users are not afraid of using the technology, nor do they fear losing their jobs to "robot instructors." For example, in a survey of teachers' reactions to educational computing, it was found that different uses of computer technology elicit different concerns about the technology. Teachers' concerns did not address a total package of technology, but rather specific applications of the technology, demonstrating that computers *per se* were not troubling to them (Wedman, 1986). So what *does* trouble potential adopters of information technologies? Even older adults, often stereotyped as technophobic, turn out to be some of the biggest supporters and users of new technologies (Gayeski, 1988).

Inhibition of Human Contact

One of the strongest themes running through descriptions of technologies that have failed is their attempt to reduce contact among people. Even when media have been proven to be effective substitutes for classroom instruction, both instructors and students report that they don't like to learn in isolation. For example, the five year project to study two-way interactive cable TV by the Carroll Instructional Television Consortium in Illinois found that: (1) students learned about the same in traditional classrooms as through cable; (2) students reacted positively to the cable instruction; but (3) a flaw was that the cable system does not allow students to get to know their classmates (Robinson and West, 1986).

There seems to be no substitute for direct human contact. Effective teleconferencing, for instance, does not cut down on face-to-face meetings as was anticipated by many managers. As Albertson (1977, p. 40) predicted, "it can be seen that even extensive use of teleconferencing would be unlikely to reduce the present overall level of travel, and indeed seems more likely to increase it." Conferees require a proportion of direct communication; if they are not previously acquainted, they report that they need to speak more redundantly, and feel that teleconferencing is not suitable for complex interpersonal tasks such as persuasion and negotiation.

People *like* to be with other people, most enjoy traveling and getting out of their offices or homes, and many do not quite trust even the most effective instructional media. Spin-off benefits accrue from group meetings, including an initiation to the organization's culture, enhanced motivation, and unplanned discussion of topics which may or may not be directly related to the instruction at hand, but may be valuable to individuals and their organization.

Disruption of Legal/Economic Status

Many applications of information technologies imply a change in the legal or economic *status quo* which may be difficult and lengthy to implement, at best, and ultimately undesirable at worst. Even when developers and adopters support new communication methods, they may not realize the long-term impact. "Technologies are never passive. They are not inert media that simply support activities which remain indifferent to them" (Malinconico, 1983, p. 111).

Kiesler (1986) maintains that technology has three orders of effects: The first order is the intended or planned effect; the second is transient effects; the third is the unintended social effects. Case in point: the telephone. The first order effect was to improve business communication, which of course, it eventually did. However, a second unintentional effect was a lack of privacy; for instance "phonies" or people who weren't really who they said they were would call up unsuspecting owners of telephones with fraudulent offers or annoying solicitations. The third order, or social effects of the phone, has been even more pervasive; this technology has made it possible for people to sustain social relationships across the miles, and has created a new mode of interpersonal interaction, as any parent of a teen-ager knows only too well.

People generally don't resist technical change—most people probably care little about the techniques and tools of communication. They resist the *social aspects* of change—the change in their human relationships (Malinconico, 1983). The more powerful the medium, the more its ability to alter the *status quo*. "Technologically-based instruction poses a threat to the base of our present system; the more comprehensive the technology, the greater the threat. When instructional technology becomes sophisticated enough to be considered an alternate, rather than a complement, to traditional instruction, it becomes a base for the design of a new educational system" (Heinich, 1985, p. 10). Media that can "take over" include televised and filmed courses, programmed instruction, CAI, and audio-tutorial methods. Media that complement instruction include overhead projector, slides, and individual films or videotapes. Heinich gives examples of how state regulations and funding models actually prevent the adoption of new methods, such as televised "open universities" and televised instruction. For example, a state will pay for teachers' salaries but not for the production costs incurred in producing a few television series which could replace several teachers.

Lack of Appropriate Designs and Information

Technology is often led by vendors rather than designers and researchers. Because of this, hardware may be sold to users who are unprepared to design effective programs or who lack the appropriate information to use it well. McLean (1985) described the information-seeking process and information sources adopted by corporate training directors when considering interactive video. Their most common sources were vendors and consultants. Trained instructional designers are often not included on teams charged with producing programs. For example, a survey by Bohlin, Evans, and Milheim (1987) found that using an instructional designer was considered least important among a list of eight processes involved in producing interactive video. Often, adopters budget only for hardware purchases, and neglect to plan for the "learning curve" in terms of costs and time needed to develop expertise in designing and producing for new media. This results in ineffective, awkward programs which do nothing to promote the wider adoption of the technology, and in fact, may send the hardware systems into a closet until someone else can be convinced of its benefits perhaps a decade later.

An example of how a lack of appropriate designs can kill a medium is the QUBE history. The Higher Education Cable Council and QUBE offered interactive cable courses but HECC colleges relied on videotapes rather than QUBE's interactive capability. High costs, low registration, and lack of appropriate designs led to low ratings (Greene, 1979). Teaching machines, too, were killed off by "exaggerated advertising claims" for programs whose developers "supported almost no research on programming" and failed to "stray from familiar formats" (Saettler, 1968, p. 263).

Even computer-based training, a modestly successful technology, has been hampered by poor examples of what it can offer. Geis (1987, p. 3)

echoes the lament of many CBT observers: "But— what a shame—these most powerful modern devices become slaves to a weak, uninformed, and archaic model of teaching." Instructional uses of new information technologies are not the only ones to suffer from a lack of adequate information about them. Salerno (1985, p. 130) reports that among the most significant factors preventing more widespread use of computers in office information systems are the "lack of knowledge at top levels about what these machine can do, or more properly, should be able to do" and "vendors' excesses in promoting their products." Schrock (1985) reported that many faculty who participated in a federally-funded instructional design project rejected the entire notion. Quotes from faculty included, "The ID process has been badly oversold."; "The consultants are like evangelists—giving a sales pitch" (p. 20).

Technology Doesn't Work Reliably

New technologies, by definition, haven't been proven and the concrete result of this is equipment and software that don't always work as expected. Whether because of equipment failure or operator error, program malfunctions become a major source of embarrassment and frustration to teachers and trainers, who often refuse to use the technology again. For example, Robinson and West (1986) found the biggest obstacle to overcome with interactive cable was "downtime" during which the system just didn't work. The use of interactive video dropped from 1985 to 1986, primarily, speculates Lee (1987) because of a disenchantment with interactive videotape systems. Many of these potentially useful tape-based configurations just didn't work reliably during production and delivery of programs. Although videodisc systems may overcome some of the awkwardness of videotape interactive systems, one wonders how readily those who dropped their tape systems will try yet another experimental medium.

Done Better by Other Media

Some media which start out on the right track to widespread adoption become derailed by another competing technology. For example, teaching machines were superceded by CBT which Geis, (1987, p. 3) says is like a "Mercedes Benz compared with goat carts." Some predict that interactive video will replace CBT once it becomes inexpensive enough, and that CD-ROM systems will replace *both* CBT and interactive video. But not all media are replaced by something newer. Videotext is an example of a medium which was less successful in communicating information than were more traditional media. Bolton (1983, p. 152) asked, "...

can videotext offer services that actually are superior to the experience of going to a library, reading from a book or newspaper, paying your bills by phone, shopping in a mall, or talking on the telephone?"

Lack of Local Production Ability

Although a broad base of generic "off-the-shelf" software is generally needed to make a new hardware system a success, increasingly organizations want to be able to produce custom materials as well. If this is difficult or expensive, potential adopters may shy away. For example, many teaching machines had software available to be bought, but producing one's own programs was often next to impossible. Because many relied on strange configurations of tape and film, the software needed to be put into its final physical form by the manufacturers of the hardware. Currently, interactive video suffers from a similar problem. Videodiscs must be mastered in one of a handful of disc-pressing plants. Additionally, the design, production, and programming aspects of this technology may be perceived as overwhelming. Because smaller firms do not (or *feel* they do not) have the ability to produce interactive video in-house, they must use outside contractors. "When independent producers are billing that creative work at contract rates, projects costs become prohibitive" (Hope Reports, 1986, p. 15).

No Standardization

Information technologies depend highly upon a base of "off-the-shelf" software and upon easy exchange of programs among users in various locations. As systems are being developed, it is common for a number of incompatible formats to emerge, each produced by manufacturers who hope that theirs will become the "standard." Teaching machines suffered from a lack of uniformity which caused problems with conversion of programs, manufacturing, and cost-effective programming (Finn and Weintraub, 1967). In the early 1970's, a number of incompatible 1/2 inch open-reel videotape systems were marketed which used tapes that physically looked the same, but couldn't be exchanged. It was not until the emergence of the EIAJ standard for these machines that they became a viable delivery medium. That lesson was learned and remembered by the same manufacturers when the 3/4 inch U-Matic standard was adopted for the next wave of videotape machines; the result of this is obvious: U-Matic machines continue to be the workhorse of instructional media and have led video to become the most widely used training delivery system. Interactive video and videotext, and to some extent, computer-based instruction,

though, suffer from a lack of standardization. Videotext systems which depend on highly attractive displays need special-purpose decoders or cards within computers, and different systems use different hardware and protocols. Interactive video, because of the number of system components involved, is one of the least standardized of media. For example, a recent survey of interactive video hardware (Miller, 1987) identified 22 videodisc players, 73 overlay/controller devices, 69 touch screens, and 43 integrated systems on the market—each incompatible with the others.

What Is Needed?

Participatory Design

In order to make new information technologies work, new design strategies are needed. The first element of a new model for information technology design is grassroots participation in the development of hardware, software, and policies for their use (Kearsley, 1984). Often new communication systems are "dropped" on organizations from upper management without having previous input from potential users. This results in a "we" and "them" perspective with regard to the ownership of the new technology.

Schrock (1985) comments on the relatively small impact that instructional technology has had on teachers and trainers. She postulates that we have been "insensitive to the role that the human element plays" (p. 16) and reports that faculty often perceive instructional technologists as "outsiders." Decisions about when and how to use new media as well as how the content should be shaped need to be formulated from the bottom-up; such a design model leads not only to more accurate and comprehensive programs, but to wider acceptance of both the programs and the technology (Gayeski, 1981). In reviewing the literature on innovation as early as the 40's and 50's, Malinconico (1983) found that the acceptability of a change is determined by how much and how well those affected by the change participated in its implementation.

The second element of a new participatory design model is an awareness of maintaining compatibility with current values and systems. Designers need to collectively be alert to developing systems which complement the social structure at large, and the corporate culture in particular, keeping in mind legal and financial considerations (Krendl, 1986). Dede (1980, 1981, 1983) argues that new information technologies in education could create many consequences for society, and that those implications are highly dependent on how they are implemented. He predicts that a new model of instruction will evolve which *combines* teachers and machines. Rather than conceiving of models through which instructional designers fabricate systems to replace teachers or trainers, we need to develop ways to enable those teachers or trainers to themselves create technologies which supplement their own roles. In fact, it appears hopeless to even attempt to replace traditional face-to-face communication. Media such as teleconferencing will not substitute for travel or meetings. "Travel and telecommunications are better seen as interrelated elements in a social context which they help create" (Albertson, 1977, p. 42).

Finally, the third element in a new participatory design model is the infusion of new techniques appropriate to the new technologies. Although some instructional design researchers and practitioners hold that standard instructional design models are useful irrespective of the medium involved, others disagree. "Existing theoretical frameworks have been conceptualized primarily for print and audio-visual media, not interactive systems. For example, existing instructional analysis and design methods do not fully exploit the instructional potential of a medium such as instructional videodisc" (Kearsley and Seidel, 1985, p. 72). Merrill (1982, p. 19) comments that "instructional design theory has not kept pace with the increased capabilities in hardware and software." Additional research and practice need to inform new instructional design models and writing techniques, especially if we are to make better use of interactive media.

Standardization

Manufacturers of new hardware systems should be mindful of the lessons learned during the development of video systems: hardware is sold by software, and the only way to establish a base of accessible and exchangeable software is to establish standardization. The constant push for new heights of technical performance must be moderated by the need for stability. Slides, overhead transparencies, and video enjoy wide use because one can find appropriate playback hardware virtually anywhere. Although computer formats have essentially narrowed down to MS-DOS (IBM), Apple II, or Macintosh, the variety of memory requirements and color graphic display components continues to make software transportability a problem. Interactive video and CD-ROM are still quite far from achieving any sort of standardization.

Local Production

Finally, new media technologies need to be able to be controlled locally. No matter how excellent and plentiful the commercial software for any given medium might be, increasingly individuals and organizations *want to create their own programming.* While some technologies like small-format video and computer graphics programs have made local production easier and cheaper, other technologies have made it difficult, expensive, or even impossible. For instance, it is not possible for users to create their own videotext messages to be shared with their on-line colleagues: these high-end systems rely on information providers, talented artists, and elaborate frame creation systems to produce their dazzling displays. Other on-line services which don't make use of such state-of-the-art graphics displays, but merely provide monochrome upper-case letters, do allow for a higher degree of participation by users, and have become more successful. Authoring systems and languages, and expert system shells are facilitating the development of interactive video and AI-based expert systems by non-programmers. As these tools become more widespread as do the computers on which these technologies run, we can expect an increase in the use of interactive and AI systems. But tools aren't the only answer. Potential developers must be trained in not only the technology but the techniques of these new media. Inexpensive and easily-created programs which are poorly designed can only hinder new technologies.

Conclusion

As educational technologists, clearly we are faced with some challenges: much of what we have developed doesn't really "work." By coming to grips with this fact, we might more eagerly explore the areas of diffusion of innovation, marketing, and organizational communication to more effectively channel our efforts. □

References

Albertson, L.A. (1977). Telecommunications as a Travel-Substitute: Some Psychological, Organizational, and Social Aspects. *Journal of Communication* 1977, *27* (2), 32-43.

Berkman, D. Instructional Television: The Medium Whose Future Has Passed? *Educational Technology*, 1976, *16* (5), 39-44.

Bohlin, R.M., Evans, A.D., and Milheim, W. Survey of Interactive Video: Current and Future Applications. Unpublished paper, Kent State University College of Education, Kent Ohio, 1987.

Bolton, T. Perceptual Factors that Influence the Adoption of Videotext Technology: Results of the Channel 2000 Field Test. *Journal of Broadcasting*, 1983, *27* (2), 141-153.

Brewer, B. Getting "It" to Happen. *CD-ROM Review,* March/April 1988, 26-30.

Brush, J.M., and Brush, D.P. *Private Television Communications: The New Directions.* Cold Spring, NY: HI Press, 1986.

Dede, C. Educational Technology: The Next Ten Years. *Instructional Innovator,* 1980, *25* (3), 17-23.

Dede, C. Educational, Social, and Ethical Implications of Technological Innovation. *Programmed Learning and Educational Technology,* 1981, *18* (4), 204-13.

Dede, C. The Likely Evolution of Computer Use in Schools. *Educational Leadership,* 1983, *41* (1), 22-24.

Eikenberg, D. Honeymoon's Over for Interactive: Time to Grow Up. *Backstage,* September 4, 1987, 1, 8, 38, 40.

Finn, J., and Weintraub, R. An Analysis of Audiovisual Machines for Individual Program Presentation. Research Memorandum Number Two. University of Southern California School of Medicine (ERIC Document Reproduction Service No. ED029486), May 1967.

Gayeski, D. When the Audience Becomes the Producer: A Model for Participatory Media Design. *Educational Technology,* 1981, *21* (6), 11-14.

Gayeski, D. Assessing Information Technologies for Older Audiences. Paper presented at the Society for Applied Learning Technology Sixth Conference on Interactive Instruction Delivery, Kissimmee, FL, February 1988.

Gayeski, D., and Williams, D.V. *Interactive Media.* Englewood Cliffs, NJ: Prentice-Hall, 1985.

Gayeski, D., and Williams, D.V. Getting into Interactive Video Using Existing Resources. *E-ITV,* August 1987, 26-29.

Geis, G. Comprehensive History of Teaching Machines. *Performance and Instruction,* July 1987, *26* (5), 3-4.

Gordon, G.N. Instructional Television: Yesterday's Magic? *Educational Technology,* 1976, *16* (5), 39-44.

Greene, A. Poor Ratings for Two-Way Television. *Change,* 1979, *11* (4), 56-57, 72.

Heinich, R. Instructional Technology and the Structure of Education. *Educational Communication and Technology Journal,* 1985, *33* (1), 9-15.

Hempstead, R.R. Push-Button Education in University of Maryland's Nonprint Media Lab. In P.J. Sleeman and D.M. Rockwell (Eds.), *Instructional Media and Technology.* Stroudsburg, PA: Dowden, Hutchinson, and Ross, Inc., 303-308.

Hope Reports. *1986 Market Trends.* Rochester, NY: Hope Reports, Inc., 1986.

Kearsley, G. *Training and Technology.* Reading, MA: Addison-Wesley, 1984.

Kearsely, G. What Makes a Computer Think It Can Teach? In D.M. Gayeski and D.V. Williams. *Interactive Media.* Englewood Cliffs, NJ: Prentice-Hall, 1985, 111-115.

Kearsley, G., and Seidel, R.J. Automation in Training and Education. *Human Factors,* 1985, *27* (1), 61-74.

Kiesler, S. The Hidden Messages in Computer Networks. *Harvard Business Review,* January/February 1986, 46-55.

Krendl, K.A. *et al.* Assessing New Instructional Technologies: Interactive Video Learning Tools. *Spectrum,* 1986, *4* (3), 3-7.

Lee, C. Where the Training Dollars Go. *Training,* 1987, *24* (10), 51-65.

Malinconico, S.M. Hearing the Resistance. *Library Journal,* 1983, *108* (2), 111-113.

McLean, L.M. Seeking Information on Interactive Video: The Information Sources and Strategies Used by Corporate Training Developers (ERIC Document Reproduction Service No. ED259719), 1985.

McNeil, D.R., and Wall, M.N. The University of Mid-America: A Personal Postscript. *Change,* 1985, *15* (4), 48-52.

Merrill, M.D. The New Component Design Theory: Instructional Design for Courseware Authoring. *Instructional Science,* 1982, *16* (1), 19-34.

Merrill, M.D. Where Is the Authoring in Authoring Systems? *Journal of Computer-Based Instruction,* 1985, *12* (4), 90-96.

Miller, R.L. Compatibility of Interactive Videodisc Systems. Falls Church, VA.: Future Systems, Inc., 1987.

Nixon, L.D. Remote Access Instructional-Learning System (RAILS). *Audiovisual Instruction,* 1970, *15* (10), 42-45.

Robinson, R.S., and West, P.C. Interactive Cable Television: An Evaluation Study. Paper presented at the annual convention of the Association for Educational Communications and Technology, Las Vegas, January 1986.

Saettler, P. *A History of Instructional Technology.* New York: McGraw-Hill, 1968.

Salerno, L.M. What Happened to the Computer Revolution? *Harvard Business Review,* 1985, *63* (6), 129-138.

Schacter, R. A Planning and Development Proposal. Unpublished paper, State University of New York at Buffalo (ERIC Document Reproduction Service No. ED104380), 1975.

Schrock, S.A. Faculty Perceptions of Instructional Development and the Success / Failure of an Instructional Development Program: A Naturalistic study. *Educational Communication and Technology Journal,* 1985, *33* (1), 16-25.

Shorenstein, S.A. Pulling the Plug on Instructional TV. *Change* 1978, *10* (10), 36-39.

Singer, I.J. At Will and at Once: The Audio-Video Dial Access Information Retrieval System. Unpublished paper, Academy for Educational Development (ERIC Document Reproduction Service No. ED039714), 1970.

Stern, C.M. Teaching the Distance Learner Using New Technology. *Journal of Educational Technology Systems*, 1987, *15* (6), 407-418.

VanWyck, W.F. Reducing Teacher Resistance to Innovation—An Updated Perspective. In P.J. Sleeman and D.M. Rockwell (Eds.), *Instructional Media and Technology*. Stroudsburg, PA: Dowden, Hutchinson, and Ross, Inc., 1976, 291-296.

Wedman, F. Educational Computing Inservice Design: Implications from Teachers' Concerns Research. Paper presented at the annual convention of the Association for Educational Communications and Technology, Las Vegas, January 1986.

Human and Quality Considerations in High-Tech Education

Peter Smith and Samuel Dunn

Considerable attention has been given in recent years to the probability of a revolution in the delivery of education made possible by new educational technologies. For example, the authors' recent article (Smith and Dunn, 1985) describes many of the innovations which promise to make significant impacts on higher education in the years ahead. On the other hand, we have discovered that very little attention has been given to those "human" quality outcomes connected with technological forms of instructional delivery. A recent literature search revealed surprisingly few papers directly addressing this important educational issue.

We believe that the emphasis in the literature on hardware capabilities of new technologies is unfortunate, for many significant concerns about the uses of educational technologies are fundamentally questions about cognitive and affective outcomes and the human element in the educational process. To put it another way, too much attention has been given to the inputs to the delivery system, with too little attention addressed to the outputs in terms of educational outcomes. We have found that colleagues often voice considerable concern that technology will inevitably lead to a lower quality of education in a "dehumanized" learning environment.

It is thought by some that the end result of extensive use of educational technologies will be seen in adults unequipped to face life in a real world which is much more complex and rich than the pseudo-world modelled by computer controlled machines. There are concerns that interaction between faculty and students will decrease. Indeed, there are even fears that faculty will be displaced entirely by computers and that colleges will be replaced by home learning centers!

We can imagine conditions and developments in which many of these fears could be realized. There

Peter Smith is Professor, School of Education, Seattle Pacific University, Seattle, Washington. Samuel Dunn is Dean of Graduate, Professional, and Continuing Studies at Seattle Pacific University.

have been experiments with educational technology which have foundered because of poor philosophy, design, and implementation. We do not believe, however, that it is necessary or even probable that the human element in education will be depreciated through the advent of educational technology. In reality, the uses of technologies are increasing so slowly with so many starts and stops that many problems are being worked out as they arise. Further, many appropriate uses of technologies have demonstrated that they can actually enhance the human element, increasing the amount of significant interaction between faculty and student and raising the overall quality of education.

The purpose of this article will be to describe several ways in which three important humanistic aspects of education can be improved through technology: (1) cognitive learning, (2) affective learning, and (3) personal interactions. Our focus will be on higher education as practiced in the United States.

Cognitive Learning

Jerome Bruner (1960) has described the process of education as one which proceeds through three phases: acquisition, transformation, and evaluation. Because many readers will be familiar with the learning outcome taxonomy described by Bloom (1956), these will be identified to illustrate the various phases of the Bruner model.

In the first of Bruner's phases, "acquisition," information is acquired by the learner. The corresponding levels of the Bloom Taxonomy are "knowledge" and "comprehension." During the acquisition phase, the focus is upon the gathering and understanding of sufficient amounts of usable information to be able to proceed to the next phase, "transformation."

In "transformation," the learner manipulates and changes information into new and different forms. In terms of the Bloom Taxonomy, operations such as "analysis," "synthesis," and "application" occur. Finally, during the phase of "evaluation," judgments are made regarding the degree to which incoming information facilitates or interferes with learning objectives. This corresponds with Bloom's "evaluation" level.

Acquisition

Let us now turn to the ways in which technology can enhance the first phase of learning, "acquisition." Essential to the learning process, and to the development of the cognitive skills that students will need to participate in the information society, is practice in retrieving, collecting, analyzing, and communicating information. We will address the acquisition question in two ways: (1) information topically delivered in class and (2) information students have usually been requested to obtain individually.

With respect to information obtained individually, students in the past have been restricted to information available in university and community libraries and from inconvenient interlibrary loan services. Thus, they have been restricted severely in both the quality and quantity of information available. This deficiency can be corrected as students take advantage of all available information through the use of new communications technologies.

We mention four sources of information: information utilities, data banks, bibliographic retrieval services, and computerized books. Information utilities, such as Compuserve, The Source, and Dow-Jones are convenient to use and are relatively inexpensive. Typically, these utilities provide news, weather and sports, stock market information, banking and purchase services, access to data banks, bibliographic retrieval services, and literally hundreds of other services. These can be obtained by telephone communication by anyone with a computer and modem.

Data banks are accessible at relatively low cost, in both time and money, to the user and greatly enhance the quality of educational programming. It is estimated that there are over 1500 data banks accessible in the U.S. Many universities have data banks containing information of special interest to local regions. Through these data banks students can access current information about census and economics and sociological studies, all critical to quality educational programming in the social sciences.

In addition to using data banks, students can retrieve information from bibliographic reference services. Through their college or community libraries, through the utilities, or by direct subscription students can access bibliographic information from services such as The Knowledge Index. This service and some others provide access to hundreds of journals. Knowledgeable users can conduct a search for just a few dollars, saving, in some cases, hundreds of hours of work in tracking down papers on topics of interest.

Starting to be available are the computerized books—the CBs. (A good name for this storage method needs to be developed.) Entire books are stored in electronic form, typically on computer disc, videodisc, or CD ROM.

Not only can information be acquired more readily through educational technology, but lessons and entire courses are increasingly available in the mediated form.

In summary, we see that the use of information technologies greatly enhances the "acquisition"

stage of the learning process by making available to the student vast amounts of useful information, inexpensively, and immediately on demand.

Transformation

The possibility of the students' transforming learning depends to a great extent upon their learning styles as they interact with the teaching styles being used. Through individualized learning, the possibilities for transformation are increased significantly. We will consider these methodologies next.

Educational technologies will influence fundamentally the learning process through "individualized learning." Here, individualized modules are used in connection with group discussion activities and individual consultations with professors. With this approach short, self-contained units or "modules" are developed and made available to the student whenever they are ready for them.

An analysis of instruction at the undergraduate level indicates that much of the work in the typical course can be broken down into sequences of relatively small topics. These topics are quite often presented or delivered by the instructor in a lecture, discussion, or demonstration format. They are frequently presented at a lower cognitive level without significant interaction between professor and student. It is often the case that the topics are presented in sequence without opportunity for a student to demonstrate mastery either to himself or to the professor. Furthermore, all students receive the same amount of instruction on each topic, regardless of background.

In these ways current lecture and demonstration methodologies are inefficient; for too many students, much of each period is dead time. With the appropriate use of technology, this standard approach can be changed to provide better opportunities for learning for the students and better use of class time.

One successful approach involves the preparation for each topic of an interactive video lesson which presents the topic and guides the student through a series of exercises to test knowledge, understanding, and the applications of the concepts presented. Taking advantage of this new medium, the approach allows students to move at their own pace through lessons.

Some professors may use the mediated lessons as supplements available to but not required of all students. The class might be taught in the usual manner but students having difficulty are pointed to interactive video lessons for additional help. At the other extreme is the professor who does not use any class time working on those topics but rather expects the students to master them outside of class, leaving class time to work with higher cognitive level concepts based on the information students gained through the lessons. In other cases the professor may not meet the class as many times as would normally be expected because the students are expected to spend more time out of class mastering the assigned topics. This approach is pursued further below in the Faculty-Student Interaction section.

Evaluation

Students may evaluate their learning most effectively if they receive rapid and meaningful feedback to their responses. Here, technology makes a significant contribution by increasing opportunities for feedback from the professor. In computer-assisted instruction, for example, students can receive fast, accurate, and helpful feedback for virtually every response they make. With simulations, new kinds of responses can be tried with immediate evaluation of their effectiveness. Technology thus allows students to evaluate their learning by receiving essential information about the effects of their responses.

With the quality and quantity of computerized lessons now available, it is the authors' assessment that the preferred method is a combination of the above approaches. That is, in most cases students should be expected to achieve mastery of topics through interactive lessons. The number of class sessions is reduced, and remaining class sessions are weighted more toward those activities associated with higher cognitive activities, such as "transformation" or "evaluation."

Affective Learning

Two areas of focus are appropriate when examining affective response to educational technology. First, specific attitudes of learners toward instructional experiences and content disciplines may be considered. Here we might ask questions such as, "Are students generally favorable to learning in environments employing educational technology?" and "Do positive attitudes generalize toward the content of the material under study?"

A second area, more general than the first, relates to the development of values. Here we might ask "Does learning through educational technology lead learners to expand and develop their values and value systems?"

We can think of several reasons why the use of educational technologies might lead to affective growth in both attitudes and values. Let us look at the matter of attitudes first. A learning environment with the potential for fostering positive attitudes may be found in computer-based instruction. Computers are often cited as being much

more patient with student errors than their human counterparts. This is particularly true when beginners make repeated mistakes. More experienced learners also find computers forgiving of error, and "safe" when they experiment with unusual or potentially dangerous responses.

When involved with well-designed computer-assisted instruction, students interact with a patient, forgiving, and "safe" teacher. Learners find themselves at little or no personal risk. Immediate feedback to response leads to rapid learning of best alternatives. All combine to make students satisfied with their learning environments and favorably disposed to continue with instruction.

Recently one of the authors conducted a review of research in one of the fastest growing areas of educational technology: computer-controlled, interactive video simulations (Smith, 1986). Most studies reported very favorable student attitudes toward training situations. In some cases, these positive attitudes even generalized to subject content as well.

Turning to the question of values development, Kohlberg's cognitive-developmental approach to moral education (1975) helps us see how mature levels of moral judgment are attained. He suggests that one must be dissatisfied with one's current level of moral judgment before progress to the next higher level can be made. In other words, one's current moral decision level must be perceived as inadequate before one will be motivated to develop higher, more sophisticated value judgments.

Educational technology can provide an excellent stimulus for growth in values. For example, through simulations, situations can be presented in which students are confronted with life-like problems resistant to solution through "lower" levels of moral judgment. Of course, these kinds of situations could be highly threatening if experienced publicly, and students might understandably be very reluctant to enter into them. However, if they could experience working through those situations within the more private, non-threatening context of a simulation, they might be more willing to face the inadequacy of their own values and begin to develop more appropriate judgments. Thus, simulations presented via high-tech delivery systems can be used as effective instructional tools for the development of values.

It is our belief that the very qualities which make technology-based learning environments effective for cognitive outcomes also make them excellent instructional modes for affective education. High-tech education can be used to stimulate the development of positive attitudes, and spur the growth of values.

Faculty-Student Interaction

When thinking about the new technologies, many faculty fear there will be significant loss of personal interaction between faculty and students if the technologies are used heavily. It is imagined that students will spend much of their time interacting with a machine, with consequent loss of high-cognitive level interaction, spontaneity, humor, body language, and the serendipity that can occur when students interact with faculty in a more traditional mode. Further, if a student spends most of his/her time working with a machine, then the significant learning which occurs through student-student interaction will be lost.

While empathizing with these concerns, and agreeing that certain elements of personal interaction may be lost, the authors believe that other elements may be significantly enhanced to the point that more personal interaction between faculty and student can occur than has been possible in the typical classroom. It is our assessment that with the appropriate use of the educational technologies, the total impact of the professor on the student's thinking and life can be considerably enhanced.

In order to describe personal interactions possible using the technologies, we first point out selected characteristics of the current classroom. First, it should be pointed out that most undergraduate instruction in the United States does not encourage student involvement. Rather, it has been criticized for encouraging student passivity (National Institute of Education, 1984). When classes are dominated by lecture and lecture-discussion, students do not have opportunities to become actively involved in their learning. What is labelled as discussion is often done at the lower cognitive levels. Most students questions merely clarify information presented. Further, a significant percentage of student questions at any cognitive level can be predicted in advance by experienced teachers. What is spontaneous to the student is most often old-hat to the professor.

In its report on the conditions of excellence in American higher education, the National Institute of Education (1984) calls for colleges and universities to promote greater involvement in learning. They recommend "the introduction and application of learning technologies so as to enhance the quality of undergraduate learning, and to foster the interaction between students and faculty and among students."

The amount and extent of student-faculty interaction in typical lecture and lecture-discussion sessions has been studied by several scholars. As a general rule, the amount is surprisingly low. Bloom (1976) cites several studies which demonstrate that college level instruction can be improved through the use of individualization strategies featuring increased student involvement.

It is now known that with appropriate use of the technologies, considerable professorial time is freed up to be used in other teaching-learning activities. Whether the time gained will actually be used to increase interaction between faculty and student will be determined by the local situation. In some institutions professors will use the time gained to do research and consulting. In other institutions there will be increased teaching loads to provide productivity gains and improved financial student-faculty ratios. The point is that institutions will have opportunity to direct time resources back to the student, thus providing for more direct faculty-student interaction.

The first area for possible gain is the most obvious: deliver factual information by machine. It is our belief that the amount of time required to deliver typical undergraduate course information by newer technologies is approximately one-third to one-half that needed for traditional delivery. Evidence appears to be mounting (coming particularly from educational programming in business and industry) that information can be delivered better, faster, and cheaper using educational technologies (Bove, 1986). The resulting savings could be used in several ways to enhance student-faculty interaction. Obviously, the time can be used for extended office hours, with open scheduling or required one-on-one appointments with students. The time formerly used in full class sessions can be devoted to small group discussions with subsets of students who need help with particular topics.

When delivery of information is relegated to out-of-class delivery, class sessions can focus on higher cognitive and affective materials. While this is not one-on-one interaction, it should be observed that only when teaching at these levels can the professor fully reveal his/her values, world view, philosophy, and perspective on the topic in question; it is only when working at these levels that the professor can reveal his or her own *self* to students. In terms of impact on students, high cognitive and affective level teaching is considerably more effective and life-changing than lower level teaching.

But there is more news here. Even some high cognitive level instruction can be achieved through chip-based technologies. Use of interactive videodisc and other technologies with sophisticated uses of artificial intelligence, graphics, branching, and logic capabilities have much potential. While high-cognitive level machine-delivered instruction is significantly more costly than low-level instruction, the costs are not prohibitive. When spread over many institutions, the costs are within reach of most colleges and universities.

The use of the educational technologies can enhance individual communication from the professor to the student, and opportunities for the professor to critique individual work. Some professors are reporting they require their students to turn in their homework on diskettes. Professors are able to enter their corrections or commentary on the disks and return them to the student. In some cases the student is allowed to hand in the successively improved homework two or three times, thus getting closer to the "perfect paper," all in the same amount of time the professor would normally spend on one paper document. The back and forth exchange necessary in any kind of written communication is substantially enhanced using this most basic application of computers, word processing.

Experiments with electronic mail are just beginning to reveal its potential to enhance faculty-student dialogue. Students from their dormitory rooms and homes can have essentially instantaneous communication with a professor, all depending, of course, on the protocols established for communication. Whether students and professors are tied to the same information utility, such as COMPUSERVE, the professor can be hundreds of miles away from the student, without loss of communicative power. Electronic mail can also enliven long-distance learning.

To summarize, appropriate use of the technologies can free the professor for more face-to-face interactions with students, can increase the instructor's impact by facilitating high cognitive and affective level instruction, and can increase the quantity and quality of written communication between professor and student.

Conclusion

In our examples it can be seen that technology has the potential to greatly enhance the quality of each phase of the educational process described by Bruner. In the view of the authors, the key to the human uses of educational technology resides in the intelligent mix of activities, the use of the technologies to amplify the quantity and quality of information available during the "acquisition" phase, and to facilitate those higher cognitive processes which make up the "transformation" and "evaluation" phases.

The writers affirm the uniquely important value of the professor in the learning process, and see technology as a liberating force, one which permits faculty to spend increasingly greater amounts of time working within those interactive, social roles and functions which are uniquely human. We do not see educational technology as a "dehumanizing" force. To the contrary, we would agree with the observation of B.F. Skinner (1968, p. 27) that through instructional technology, a teacher begins to function: "not in lieu of a cheap machine, but through intellectual cultural, and emotional contacts of that distinctive sort which testify to her status as a human being." ☐

References and Suggested Readings

Bloom, B.S. *Human Characteristics and School Learning*. New York: McGraw-Hill Book Company, 1976.
Bloom, B.S., Engelhart, M., Furst, E., Hill, W., and Krathwohl, D. *Taxonomy of Educational Objectives: Handbook I, the Cognitive Domain*. New York: Longman, 1956.
Bove, R. Video Training: The State of the Industry. *Training and Development Journal*, 1986, *40*(8), 27-28.
Bruner, J.S. Readiness for Learning. *The Process of Education*. Cambridge: Harvard University Press, 1960.
Carlson, W.L. Integrating Computing into the Liberal Arts: A Case History. *T.H.E. Journal*, 1985, *13*(2), 95-100.
Driscoll, J.P. Dehumanize at Your Own Risk. *Educational Technology*, 1978, *18*(6), 4-36.
Karwin, T.J., Landesman, E.M., and Henderson, R.W. Applying Cognitive Science and Interactive Videodisc Technology to Precalculus Mathematics Learning Modules. *T.H.E. Journal*, 1985, *13*(1), 57-63.
Kohlbert, L. The Cognitive-Developmental Approach to Moral Education. *Phi Delta Kappan*, 1975, *56*(10), 670-677.
Morrisett, L.N. Technology, Humanism, and Higher Education. *Educational Researcher*, 1974, *3*(8), 15-16.
National Institute of Education. *Involvement in Learning: Realizing the Potential of American Higher Education*. Final report of the Study Group on the Conditions of Excellence in American Higher Education. Washington DC: National Institute of Education, 1984.
Smith, P., and Dunn, S. Tomorrow's University: Serving the Information Society—Getting Ready for the Year 2000. *Educational Technology*, July 1985, *25*(7), 5-11.
Smith, P. Instructional Simulation; Research, Theory, and a Case Study. *Proceedings of Selected Research Paper Presentations*. Annual Convention of the Association for Educational Communication and Technology, 1986. Washington, D.C.: AECT, 1985.
Skinner, B.F. *The Technology of Teaching*. New York: Appleton-Century-Crofts, 1968.

Distance-Delivered Instruction: Making It Work

Barry Willis

When developing and delivering distance education, the instructor is well advised to be familiar with the challenges and opportunities to be encountered throughout the process. Too often, promoters of distance delivered instruction minimize the planning, time, and effort it takes to learn effectively via distance education.

In a "traditional" classroom setting, the instructor (and students) are privy, on both a conscious and subconscious level, to various forms of input and feedback not readily available in a distance education setting.

Subconscious teacher/student cues may be as subtle as a wandering eye indicating a lack of attention or a stifled yawn while presenting a supposedly critical point. In fact, one factor that separates the "good" teacher from the truly exceptional teacher is his/her ability to constantly remold the critical instructional points being presented to fit the situation at hand. The art here is not so much changing the actual content being presented, but in capitalizing on the ever-changing teacher/student dynamics to insure that critical instructional points are presented (and received) with maximum clarity.

While attainment of this goal is elusive under any circumstance, it becomes even more challenging in a distance delivery setting. In many distance education courses, for example, the instructor has had no personal contact with the student prior to the course. As a result, the "grapevine" that typically informs students of an instructor's strengths, weaknesses, and personal characteristics is either limited or non-existent. In addition, the instructor/student relationship often lacks the unifying realm of experience, either on a personal or academic basis, that results when teachers and students alike are a part of the same social or geographic community.

Research indicates that even when interactive technologies are used, such as audio conferencing

or two-way television, the dynamics that result are different from those encountered in more traditional classroom settings. Despite the new challenges posed by teaching at a distance, there are a number of strategies that the distance educator can use to build teacher/student linkages and improve instructional effectiveness.

Strategies to Improve Instructional Effectiveness

1. Get to know your students (and let them get to know you) early on in the course or even in a presession event before the instruction begins. Exchange photographs, mini-biographies, descriptions of backgrounds, personal interests, and anything else that will help you and your students understand each other better as people.

2. Treat distant students as if they were in the same room by being warm, polite, and responsive.

3. Keep your instructional program flexible by offering a choice of delivery methods, time frames, and post-presentation activities to maximize student learning.

4. Emphasize early on that you and the students are part of the same distance education "team." It is the responsibility of both you and your students to see that instructional goals are met. This will require opportunities for group and individual input and feedback in an environment that encourages open communication.

5. When developing a distance delivered course, make sure that the "context" as well as the content is relevant to your students. By definition, distance education courses typically bring together students from diverse geographical and often cultural backgrounds. Make sure the contextual examples you use don't obfuscate the content you are presenting.

6. Be aware of and respect cultural differences in communication patterns. Don't assume that immediate responses to questions/inquiries indicate enhanced comprehension. It is very apparent in an interactive distance education course when a student, or the teacher for that matter, is ill prepared and is attempting to "buy time" until a correct response can be formulated.

7. If at all possible visit and teach class from each site one or more times during the course.

8. Provide opportunities and encourage students to use available technology to work among themselves. Consider joint presentations in which students from different geographic and/or cultural settings look at specific issues from their unique perspectives. Often students participating in distance education courses are isolated from other students. Anything that you can do to bring them closer to their fellow students will make their distance education experience more enjoyable.

Barry Willis is Associate Chancellor, Distance Education/Academic Planning, University of Alaska, Anchorage.

9. As teacher, strive to feel comfortable in the role of "skilled facilitator" as well as "content provider." A primary goal in most distance education courses is to make all students feel they are working together towards some definable instructional goal. Towards this end, your role as facilitator is critical.

10. Whenever and wherever technology is used, technical problems will occur. Don't be embarrassed or defensive when this happens. Work with students and technical staff to minimize problems before they occur, and just as importantly, have instructional contingencies planned for the inevitable "glitch" when it happens.

11. Even in your "lectures" make sure you build in plenty of opportunities for discussion and interaction. Effective distance delivered instruction requires the full participation of both teachers and students.

12. Break lectures into small content blocks interspersed with interactive activities.

13. Begin each class with a statement of purpose or objectives.

14. Follow an outline that has also been provided to the students. Periodically refer to your place in the course's organizational scheme.

15. When teaching in an "interactive" mode (e.g., audio-conferencing), give students enough time to respond to the questions you pose. Don't be afraid of silence.

16. If you are having difficulty getting students from some sites to respond, don't hesitate to call on specific students by name.

17. As teacher it is your job to firmly control "verbal traffic" by regulating which sites (and students) have the "floor."

18. While in the midst of course planning, realize that it will typically take longer to present a course lesson in a non-traditional as opposed to a traditional teaching format. Don't attempt to move through the content with unrealistic speed at the expense of student comprehension.

19. Be yourself and relax. It makes your students feel more comfortable and the course more enjoyable.

Most importantly, don't feel obligated to re-create "traditional" instructional methods in a "non-traditional" distance education setting. Rather, minimize the difficulties inherent in distance education through effective planning, and constant feedback, while enjoying the unique opportunities that distance-education offers. □

Distance Education and Student Procrastination

Thomas W. Wilkinson and Thomas M. Sherman

The old proverb, "Nothing makes a person more productive than the last minute," probably strikes a familiar chord in most. However, it seems to be far from an accurate description of a significant proportion of students in distance education programs. Estimates of noncompletion in distance education range from 30 to 70 percent. While many factors may contribute to noncompletion, clearly our understanding of the reasons why people fail to complete distance education courses is far from complete or satisfying. At least one explanation for why students fail to finish what they begin is procrastination.

Bliss (1983) defined procrastination as needlessly postponing tasks. Unfortunately, distance education appears to provide a fertile ground for putting off responsibilities. Distance education is generally characterized by the separation of students and instructors during the primary modes of education delivery and focuses on independent self-directed learning (Evans, 1986; Feasley, 1983). By its very nature, distance education allows students to control much of their own learning including decisions about when to study and how to respond to academic assignments. We were somewhat surprised as we began looking into procrastination in distance education to find that relatively little information on this problem was available. Because our own experience told us that procrastination was likely to be a major factor in noncompletion by students, our curiosity was aroused regarding the extent to which others experienced similar perceptions. We were also interested in identifying strategies by which procrastination may be reduced. Here, we report the results of an exploratory investigation in which we probed distance education administrators and professors' opinions and actions regarding procrastination.

Thomas W. Wilkinson is Director, Learning Resources, New River Community College, Dublin, Virginia. Thomas M. Sherman is Professor, Division of Curriculum and Instruction, College of Education, Virginia Polytechnic Institute and State University, Blacksburg, Virginia.

Methodology

To determine the extent to which procrastination was a problem, we interviewed the administrators and professors of two relatively diverse distance education programs. These interviews relied on open-ended questions and follow-up probes to elicit information in four areas:

1. General demographic information about the individuals such as degrees received, courses taught, and number of years teaching distance education courses.
2. General perceptions about the distance education program in which they were involved, including strengths and weaknesses of distance education as an instructional delivery strategy, program structure, and the advantages and problems for students, faculty, and administrators.
3. Perceived reasons for student attrition and noncompletion and what was being done about these issues; and
4. Their definitions of the concept of procrastination as a cause for noncompletion.

Results

The first program was located in a rural portion of a southeastern state. This program falls under the classification Giltrow (1989) called "distance education coordinating office within a conventional institution." It is a relatively small program involving approximately 200 students and seven to nine classes per semester. Courses offered include architecture, psychology, history, business, office technology, and health. A variety of distance education strategies are employed, including cable television, video store rental, and viewing in the college's independent learning lab on the campus. Faculty are assigned to all classes and are available to consult with students via mail, telephone, and in person by appointment. Students work independently of the faculty on their coursework to obtain course content by videotape, textbooks, supplemental readings, and workbooks. The distance education coordinator of this program has been involved in the program from its beginning. She has six years of experience working with independent learning programs. The program offices are located in the independent learning lab of the learning resource center. Program staff typically have contact with students and answer questions when possible or relay messages to faculty.

The three faculty interviewed for this study have from one to six years experience teaching distance education courses. Two of the faculty are full-time with over thirty years combined teaching experience. The third faculty member is a full-time member of the college's independent learning lab and also teaches a course on microcomputers via distance education.

The second distance education program is located in a densely populated urban area of the same state, and has been in operation over 15 years. According to Giltrow (1989), it can be classified as "an institute within a conventional institution." It is a large operation with over 3,100 students enrolled in over 50 courses per semester. The program offers instruction in a wide variety of subject matter, including geology, English, economics, and psychology. This program is heavily print-based, although approximately one fifth of the courses are delivered on video, which can be viewed on cable television or at learning labs on the college's five campuses. Faculty are assigned to all classes; contact between student and faculty is generally by mail or telephone. Students work independently of the faculty.

The distance program director had been in his job only five months at the time of the interview; both he and his predecessor, who directs the college's total distance education and instructional technology efforts, were interviewed. These administrators had over thirty years of combined experience in educational course delivery. Program offices are located separately from any of the five campuses the distance education program serves. Though the program director and staff rarely see students, they do relay telephone messages from students to faculty. Two faculty members in this program were interviewed individually. One of the professors had taught distance education courses for the last seven years; the other had been teaching a business course by distance education for five years. Both held master's degrees and were pursuing doctoral studies.

Finally, we examined curriculum materials that were distributed to students, which included course materials and syllabi, check sheets, program descriptions, televised course schedules, and learning lab hours. We also reviewed reminder letters which were periodically mailed to students.

In our interviews with these eight distance educators, four themes consistently emerged. Below, we identify these themes, discuss their implications for distance education programs, and describe the potential role procrastination may play in student noncompletion.

1. Distance educators are concerned about the problems of noncompletion in distance education courses.

All of the distance educators interviewed expressed concern about high noncompletion rates. While the program directors had more specific data on noncompleters and nonstarters (30 to 45 percent), distance education faculty expressed a concern that the numbers were "too high." When asked what they thought the biggest problem in distance education was, all distance educators responded unanimously that it was the high number of dropouts. Thus, it appears noncompletion and dropout are perceived as major problems. This was not surprising because attrition and noncompletion have received more attention than any other area of distance education research (Garrison, 1987; Rekkedal, 1983). Directors of both programs indicated that their attrition rates were in line with other programs with which they were familiar and appeared consistent with the literature as well (Feasley, 1983). In general, the directors appeared to consider noncompletion as a common cost of distance education programming. On the other hand, distance education faculty appeared more concerned over the number of students who failed to complete the courses. During interviews, they became more animated and reflective when searching for reasons to explain their students' failure to finish coursework.

2. Explanations for student noncompletion.

All eight distance educators found noncompletion perplexing, but, nonetheless, they were able to suggest reasons for student noncompletion. Most of these reasons were based on experience, although distance education program directors indicated a knowledge of current research in this area. For example, the directors had recently either initiated or planned to initiate new pacing procedures to help reduce noncompletion in their programs. Several of the reasons given for noncompletion appeared in nearly every educator's explanation. These were: (1) students do not have realistic expectations about what is demanded in a distance education course; (2) students need more structure; (3) students who finish generally begin early and pace themselves; and (4) students in distance education are really atypical—they have many other responsibilities which may take priority over their coursework. In general, all of those interviewed expressed an opinion similar to a statement by one of the distance educators that "distance education isn't for everyone." This perception seemed to be based on the belief that intrinsic in the characteristics of distance education are certain factors which may not provide an environment in which all students can be successful.

While the faculty did not consider themselves up-to-date on the distance education literature, their ideas clearly coincided with the ideas of the distance education program directors and with many of the ideas expressed in the literature (e.g., Harter, 1969; Gatz, 1986). Unfortunately, mixed success is reported for many of these strategies to combat student noncompletion. For example, measures such as increased pacing and use of contracts have at times proved successful while at other times unsuccessful. Pacing and increased structure were the strategies most appealing to the distance educators interviewed for this study.

3. Lack of knowledge and information about procrastination.

When the topic of procrastination was explicitly introduced, invariably the distance educators would pause and almost appear surprised by the question. In fact, most had trouble defining and discussing the issue of procrastination. However, there was a fairly common perception of what constituted procrastination following some probes on the issue. In general, this can be characterized as "putting off today what you can do tomorrow." Additional questioning on the issue of procrastination provided little insight. Though their conceptions of procrastination generally suggested time management and organization problems, they had little insight into the potential causes of procrastination. One professor did suggest that procrastination might be a "symptom for something else," such as the fear of failure or the fear of success, an opinion which is consistent with the current literature on procrastination. Though not extensive, the research which does exist appears to indicate that procrastination is a complex interaction of behavioral, cognitive, and affective components (Solomon and Rothblum, 1984; Rothblum, Solomon, and Murakami, 1986; Burka and Yuen, 1983).

None of the distance educators had any specific information from students on procrastination. The curriculum and orientation materials, while stressing the point that it is important to begin early and maintain a steady rate of work, did not emphasize procrastination as a potential problem. In general, it appeared the major advice given to distance education students in the syllabi and other literature focused on time management. Thus, it appears students and distance educators may have a relatively simple and unformed concept of procrastination.

4. Lack of time to address the problem of procrastination despite the intent to do so.

One of the most interesting themes emerging from this study was the extent to which distance educators engaged in some of the same behaviors as their students. All of the distance educators indicated they wanted to do more about reducing student noncompletion and procrastination; however, they cited lack of time, too many other commitments, and lack of energy as reasons why they were unable to do more. Of course, these are some of the same reasons that these faculty reported students telling them as excuses for not completing their courses. As one faculty member said, "I identify with this (procrastination) so much; I have good intentions but. . . ." One program director even showed us a pile of printouts eight inches high which contained data on distance education courses from the previous semester. He indicated he just simply had not had time to analyze the data and did not know when he would be able to because of other priorities. There was a clear sense of frustration on his face.

Conclusion

It appears that procrastination is a recognizable phenomenon which potentially may impact negatively on the success of distance education programs. These faculty and administrators certainly appeared to recognize the existence of procrastination, and the literature supports that noncompletion is a problem. The main issues with procrastination appear to be a failure to understand the complex nature of procrastination as well as a lack of effective strategies to combat procrastination. Studies of procrastination in traditional higher education settings (e.g., Semb, Glick and Spencer, 1979) tend to indicate that developing structure which forces students to initiate study activity is a major factor in combating procrastination. However, implementation of strategies which increase the structure of distance education may reduce procrastination, but, at the same time, eliminate many of its positive characteristics. Thus, distance educators need to define more clearly and investigate procrastination as a phenomenon within the context of distance education itself. It appears there is now sufficient reason to believe that these investigations should be put off no longer. □

References

Bliss, E.C. *Doing It Now*. New York: Charles Scribner's Sons, 1983.

Burka, J.B., and Yeun, L.M. *Procrastination: Why You Do It, What to Do About it*. Reading, MA: Addison-Wesley, 1983.

Evans, A. Media Managers and Distance Education. *Media Management Journal*, 1986, 5, 22-23.

Feasley, C.E. Serving Learners at a Distance: A Guide to Program Practices. ASHE-ERIC Higher Education in Research Report No. 5 (ED 238 350), 1983.

Garrison, D.R. Researching Dropout in Distance Education. *Distance Education*, 1987, 8, 95-101.

Gatz, F. Personal, Instructional and Environmental Factors Associated with Completion and Attrition in Correspondence Study and Distance Education. Unpublished Doctoral Dissertation, Indiana University, Bloomington, IN, 1985.

Giltrow, D. *Distance Education*. Topical paper presented at

Association for Educational Communications and Technology Conference, Dallas, TX, 1989.

Harter, D. *Why SUNY Students Fail to Complete Independent Study Courses.* State University of New York, (ERIC Document Reproduction Service No ED 035814), 1969.

Rekkedal, T. The Dropout Problem and What to Do About It. In M.A. Daniel, J.R. Stroud, and J.R. Thompson (Eds.), *Learning at a Distance: A World Perspective.* Edmonton, Canada: Athabasca University, International Council for Correspondence Education, 1982.

Rothblum, E.D., Solomon, L.J., and Murakami, S. Affective, Cognitive, and Behavioral Differences Between High and Low Procrastinator. *Journal of Counseling Psychology*, 1986, *33*, 387-394.

Semb, G., Glick, D.M., and Spencer, R.E. Student Withdrawals and Delayed Work Patterns in Self-Paced Psychology Courses. *Teaching of Psychology*, 1979, *6*, 23-25.

Solomon, L.J., and Rothblum, E.D. Academic Procrastination: Frequency and Cognitive-Behavior Correlates. *Journal of Counseling Psychology*, 1984, *31*, 503-509.

Distance Learning and Public School Finance

Brian D. Monahan and Charles Wimber

The time to integrate technology into education in the form of distance learning has come. Hundreds of school districts now operate computer bulletin boards for everything from PTA notices to cross-county communication.

Affordable Standards

A combination of new technology and new demands upon public education has made distance learning a real possibility. Although the technology is constantly changing, we have reached the point at which there are standards on which most school districts can agree. These standards will make it possible for information to be shared in ways that were never possible before. The "de facto" standards for distance learning include the following:

- ASCII TEXT for text messaging
- VHS VIDEO for video messaging
- DIGITAL AUDIO for audio messaging
- TOUCHTONE for voice messaging

As a result, diskettes, videocassettes, audio cassettes, and voice messages can be the components of what is called "lessonware." The skills required for the creation of lessonware include the ability to play, record, dub, send, receive, and update information created with the integration of voice, audio-video, and texts. Educators who wish to create lessonware will require a "Desktop Video Production Workstation." (See Figure 1) That workstation will be equipped with the following:

- an AM/FM stereo with wireless mike
- a TV/VCR recorder/camcorder/switcher
- a phone with a telephone answer device
- a personal computer system

The personal computer system will have word processing and storage capabilities, a terminal software program, a BBS system, and assorted

Brian D. Monahan is Associate Professor of Computer Science and Coordinator of Graduate Programs in Computing at Iona College, New Rochelle, New York. Charles Wimber is a data communications consultant in Denver, Colorado.

other utility software to improve the quality of the lessonware produced and make that production more efficient. The desktop video production workstation will have the capability of integrating text, images, audio, and voice onto a videocassette.

Such workstations, located in classrooms that are equipped with two-way, reverse passes to a cable operator's head-in can upstream lessonware to a public broadcasting station (PBS) where the lessonware can be prerecorded. After midnight, the PBS can broadcast the lessonware to be captured by programmable VCRs, either from an over-the-air cable systems or from direct satellite. This after-midnight technique makes a more efficient use of a PBS's license. Obviously, for true education to take place, mechanisms must exist for responses to the lessonware that is broadcast. Such responses can occur using both voice messaging, either directly by telephone or via a telephone answering device, or through the E-mail, using a dedicated bulletin board system.

Implications for School Finance

What is the relationship between distance learning, as it is described in this article, and public school finance? Public school finance acts passed by state legislatures and enforced by state departments of education are coming under increasing pressure from property tax payers and from students and their parents. In Colorado, for example, the State Board of Education and General Assembly are facing such pressure from a "class action suit" called *Nancy Hafer v. Colorado State Board of Education*. The Hafer Case has two sets of plaintiffs: school children and taxpayers. The school children's complaints will deal with equal educational opportunity; the taxpayer complaints will deal with taxing equality.

The downstream/upstream distance learning system described in this article is one way to work toward geographic equity in terms of access to educational opportunity. That geographic equity can become a reality due to technology. Of course, start-up costs for a program such as the one described here can be significant. The districts least able to provide a variety of traditional educational opportunities because of their size, demographics or location would be least likely to have the resources to take advantage of the technology described here. A possible solution is an inter-public school voucher system that could be financed by a recapture school finance act. That act would pool property taxes, sales taxes, income taxes, and other taxes which go to support education; each district would recapture portions of those taxes based on a combination of weighted needs and student population. (Many state school financ-

Figure 1

Self-Contained MessageWare Workstation
(Close-Up as a Cabinet)

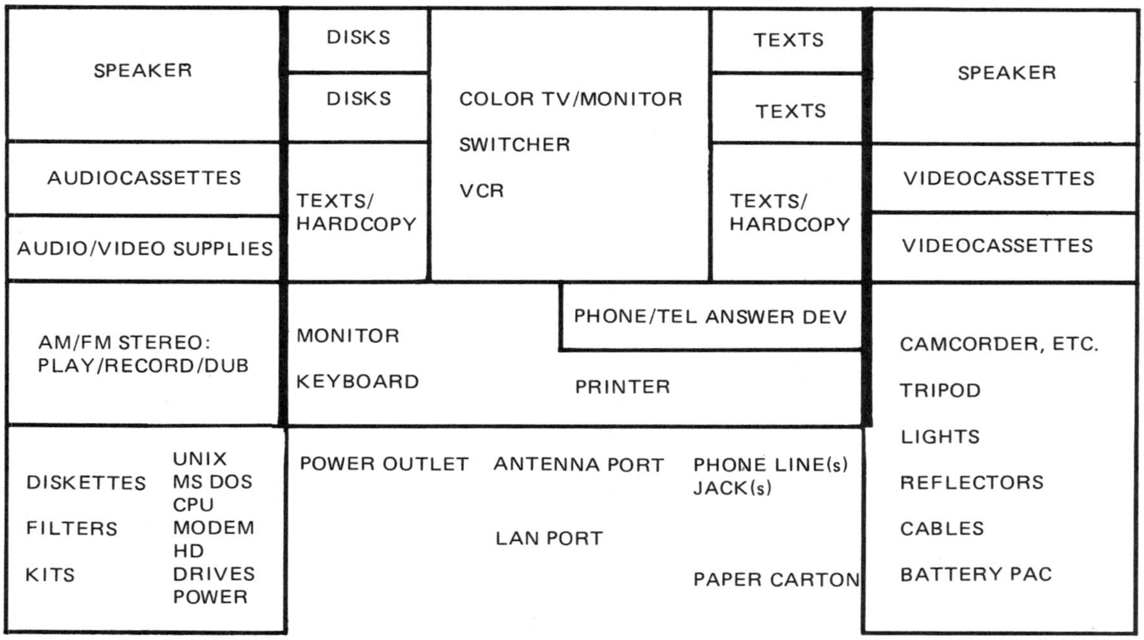

SPEAKER	DISKS	COLOR TV/MONITOR	TEXTS	SPEAKER
	DISKS		TEXTS	
AUDIOCASSETTES	TEXTS/ HARDCOPY	SWITCHER	TEXTS/ HARDCOPY	VIDEOCASSETTES
AUDIO/VIDEO SUPPLIES		VCR		VIDEOCASSETTES
AM/FM STEREO: PLAY/RECORD/DUB	MONITOR	PHONE/TEL ANSWER DEV		CAMCORDER, ETC.
	KEYBOARD	PRINTER		TRIPOD
				LIGHTS
DISKETTES — UNIX MS DOS CPU	POWER OUTLET ANTENNA PORT PHONE LINE(s) JACK(s)			REFLECTORS
FILTERS — MODEM HD	LAN PORT			CABLES
KITS — DRIVES POWER	PAPER CARTON			BATTERY PAC

ing systems already work, at least in part, in this way.)

Distance learning can make inter-public-school voucher systems affordable, efficient, effective, and equitable—if the downstream/upstream operating system is standardized and horizontally integrated for voice, audio, video, and text messaging. Distance learning has a teacher with students going to remote study groups electronically with video, audio, E-mail, or voice messaging.

The benefits to students go beyond the traditional subject matter that will be covered in the "distance-learning classrooms." PLAY, RECORD, DUB, SEND, RECEIVE, CREATE, AND UPDATE are the absolutes to survive and compete in a world-wide, information-driven, flexible production economy. Students are not the only ones who need these skills to survive. Distance learning can also be used by adults who need to be retrained to participate in an information economy.

An inter-public school system that takes advantage of distance learning will give those teachers who are involved with distance learning a competitive edge. That edge may result in the creation of new types of magnet programs. Traditionally, magnet programs have been special programs that were so attractive that students were willing to travel, sometimes long distances, to participate in them. These programs had to maintain excellence to survive. Magnet programs that apply distance learning provide a way to share excellence, with the travel component being eliminated.

"Voucher" is a word with the worst type of connotation for those who support public education. The type of vouchering that should be involved with distance learning will be quite different. If teachers do not lobby for the technology and training to participate, it is possible that they may be bypassed by private and home schooling organizations that do have such access and that can deliver the lessonware more reasonably, especially if they are permitted to do so by using non-certified teachers for all or part of lesson production.

Some Examples
Relatively small-scale examples of the system

being described in this article already exist. In San Antonio, Texas, TI-IN uses a "master-slave, forced-migration" scheme to export lessonware from San Antonio to isolated schools in the rural areas. CompuServe Information Service, through databases like ASK EINSTEIN, can assist distance learning. Special SIGs already exist for educators with a variety of interests. More could be formed to offer support for teachers doing distance learning.

Do we want distance learning to survive and flourish? Our research and experience suggests that we have no choice. For it to succeed in the best possible way, we need coexistence between public, private, and home schooling where there can be peer relationships through on-line activities on the part of teachers, parents, students, and even principals and superintendents.

What is the largest industry in the U.S.? It is K-20 education. It is the creativity and innovation of classroom teachers that will determine if our country has the "knowledge-based" work force necessary to compete. With VHS as a video standard, hi-bias as an audio standard, touchtone as a voice standard, and ASCII files as a text standard, distance learning can be horizontally integrated. "Play, record, dub, create, send, receive, and update" become absolutes that must be learned by teachers, administrators, and their student populations (K-20 students and adults). Professional educators who can direct the production of lessonware become the "entrepreneurs" that our nation needs in order to compete. Lessonware that is produced by these master teachers for magnet programs can become a shared resource at the local, state, and national levels.

The Future of Distance Learning

The process will not end in schools. Lessonware can lead to "ECONOWARE" for business, "HEALTHWARE" for doctors and hospitals, "SAFETYWARE" for police and fire departments, as well as applications for virtually any segment of American life or society. Eventually, the technology can become so widespread that lessonware becomes "messageware" that adds to creativity and innovation, as well as to the quality of life. Messageware can become "thoughtware" that can give the nation the competitive edge it is now seeking. □

An Information Technologies Workstation for Schools and Homes: Proximate, Border Zone, and Distant Educational Possibilities for the Future

J. Allen Watson, Sandra L. Calvert,
and Rosann Collins

Currently among the nation's general populace there seems to be a recognition of a relatively "quiet revolution" going on around them. For those who are eager supporters of the new information technologies, this vision of a "quiet revolution" with its accompanying attitudes of "it's still business as usual" is extremely troubling, especially given the large number of public educators who seem to hold the "status quo" viewpoint (Watson, Calvert, and Brinkley, 1987). Lethargy is abhorrent to those scientists who laboriously collect and catalog the runaway pace of this "not-so-quiet" revolution. We believe that technological change is far beyond any questions related to "business as usual."

The Information Technology Revolution, although more developed in some social contexts (business) than others (public schools), has matured to the point that questions of equity (who benefits, when, and how) and control (who plans and directs) are currently paramount. The old educational arena is fragmenting, and a new educational pattern is emerging: proximate to distant educational sources (Stephenson and deLandsheere, 1985).

We are concerned herein with questions of equity and control in plans for the individual learner (e.g., a ninth grade student) in both the public school and home contexts. We have searched the literature to discover what proximate/distant education is and what its implications are for our hypothetical ninth grader. We also propose a possible proximate/distant educational model, a skeleton of what might be, predicated on what is

The authors are with the Children and Technology Project, Department of Child Development and Family Relations, University of North Carolina, Greensboro, North Carolina.

now available, while recognizing humankind's limitations in predicting the future. We project a scenario whose importance lies in its potential heuristics. We encourage colleagues to develop bottom-up as well as top-down technological applications and to promote public supported ventures where less empowered user groups (public school personnel and families) are unlikely to attract business solutions.

To highlight the user groups' problem, we cite the example of a scientist and the significant milestone reached in developing a scientific workstation (Crecine, 1986) connected to a small local area network (LAN) which is interneted with the National Academic Research Network (Jennings, Landweber, Fuchs, Farber, and Adrion, 1986). Today, all scientists depend upon information technologies: i.e., computers, networks, databases, etc., and are said to be handicapped without them (Jennings et al., 1986). What are their LAN workstations predicted to be?

Crecine (1986) says that the new scientific workstation will be available within the next two years. The workstation will utilize UNIX-based systems software, be five to ten times more powerful (execute three million plus instructions per second with two million plus bytes active memory and 30 million plus bytes hard disk storage) than current microcomputers, and provide the user with the most advanced design aids, graphic tools, and knowledge-based systems.

Access to both local area and national networks has become a major issue as scientists discovered that "communication capabilities were more important than computation." The distinguishing trait of modern or future scientists may well shift from the scientist in the laboratory to the scientist at the computer workstation in the laboratory.

As the information technologies filter down to our ninth grader, we envision the same handicapping condition to exist. Education is shifting paradigms from how much or what a person knows to what a person knows plus the ability to re-organize and structure mental functioning with the aid of new cognitive and educational technological tools (Pea, 1985). If proximate and distant work is so important to the success of the scientist, businessman, politician, doctor, and manufacturer, then proximate and distant education will take on new meaning as well for the educator and student.

What is proximate (near) and distant (far) education? What does a proximate border zone distant educational model have to say about the future of education? We believe that the public school institution of the industrial age has served us well, but it is undergoing radical change. Our social institutions are incorporating the information technologies, and many will profoundly change. Public schools, colleges, and universities are slowly changing, will begin to change more rapidly, and may be transformed to information age institutions in the near future.

Information age public schools, colleges, and universities will be characterized more by distant education, just as today's educational institutions are characterized by "the classroom and teacher" (near education). Proximate learning means education in a local setting (teacher or parent plus learner) where materials (information) are maintained, selected, and presented. Distant education is learning which comes from afar and is exemplified by satellite, television, network, and database technologies applications.

What will happen when and if learning becomes truly individualized (e.g., where educational packaging matches cognitive styles) and predominantly self-directed, with few geographic or academic level constraints? We don't know, but such activity can be envisioned with a student workstation which connects the local classroom or home with the distant world.

In order to explain our proximate ↔ border zone ↔ distant educational model, we use components of the following example concerning our hypothetical ninth grader. We will describe a scenario which takes place in a student workstation in the future.

"Susie Anybody," our ficitious ninth-grade student in the year 2000, is writing a paper about embryonic development and its effects upon the mother. Susie's teacher directs her to the local school encyclopedic data base on this subject, which is stored on compact disc and includes audio, video, written, and graphic information which she calls up on her workstation. Susie also watches a videodisc about the topic and asks many questions after the background information has been presented. Susie wants additional information, since some of her questions failed to be answered by her school's data archives. She decides that she needs up-to-date answers and activates the switch which allows her LAN (Local Area Network) to interface with the border zone data integration source, the regional university, which is 25 miles away. During the past 15 years this university has become the region's information manager and has the special task of integrating the local public's and

school's informational needs and demands with local/national data sources. General databases have multiplied and branched to the point that what is known about any subject has a database name (including expert systems and intelligent tutors) and is locally organized and maintained by the regional university on its supercomputer mainframes.

Susie instructs her computer to contact the regional data source on newly pregnant mothers. She is connected with the university's Department of Child Development, whose professors are responsible for maintaining, organizing, and generating new information (research) for students and the general public in their area of specialty. Susie transmits her questions via computer and receives some specific electronic answers from a Child Development professor, and an intelligent tutorial package responsive to her age and cognitive style. Up-to-date national medical and social information accumulated and stored about pregnancy is transmitted to Susie from supercomputer to supercomputer to her workstation. Susie gets her answers quickly and in as much detail as she desires (video, audio, written, and numeric). She then begins to write (word-process) her paper, interspersing edited pieces of video to visually emphasize her verbal points.

A Learner Workstation

The learner of tomorrow will need access to all information technologies, and the best approach to insure a common source of connection seems to be the workstation. Credit is given to Xerox Corporation's Palo Alto Research Center (PARC) for the initial 1979 design of an integrated, LAN microcomputer workstation. Since then. many computer manufacturers have developed their versions of a workstation. We have designed a learner workstation (see Figure 1) which includes not only the depicted devices, but incorporates the hardware specifications as described by Crecine (1986).

We envision a world where such workstations will become commonplace in home, school, and business environments. A workstation, as depicted in Figure 1, is an integrated technological learning environment which supplements teacher (parent) to student interactions. We maintain that the best way for a human to learn is from another human and that electronic learning will always be secondary, but of increasing significance. The more

Figure 1

The Learner Workstation

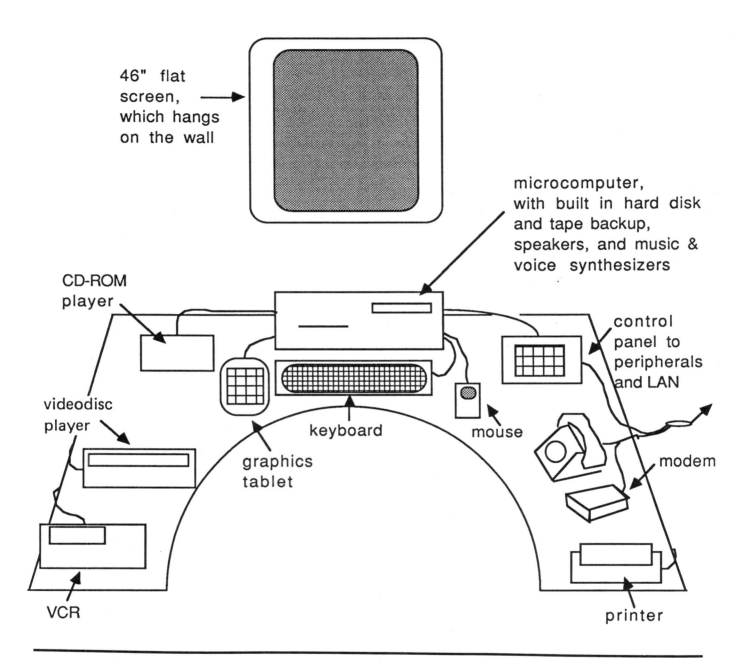

46" flat screen, which hangs on the wall

microcomputer, with built in hard disk and tape backup, speakers, and music & voice synthesizers

CD-ROM player

control panel to peripherals and LAN

videodisc player

keyboard

mouse

graphics tablet

modem

VCR

printer

interactive and varied the workstation, the more humanlike the learning experience will be.

A Proximate ↔ Border Zone ↔ Distant Educational Model

The workstation becomes a very powerful learning environment when it is placed in a proximate ↔ border zone ↔ distant educational context, as depicted in Figure 2. The workstation, located at home or school, is envisioned as the point at which both proximate and distant learning begins and ends. The electronic input-central processing-output devices pictured in Figure 1 are the tools to learn from technological systems.

The border zones are the boundaries between proximate and distant educational technological

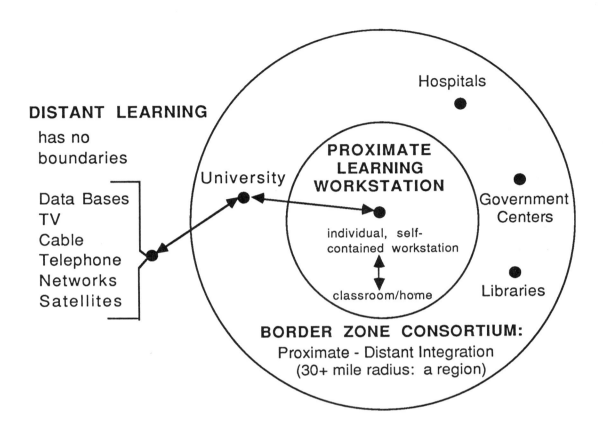

Figure 2

Proximate ↔ Border Zone ↔ Distant Educational Model

systems. The border zone in our model contains a greatly expanded version of a local network. We think that public schools and homes will be interneted into a "local-wide area network" which includes regional universities, hospitals, government centers, and libraries. The border zone networks will be the point of interaction between local and national/international systems. The border zone will be the point at which major problems are addressed concerning an infrastructure for massive data storage, a data flow system, and a hierarchy of interneted networked systems.

The distant learning component has no boundaries and is characterized by national and international communications. Increasingly, we will become global learners as learning technologies foster the development of academic specialties through distant education.

Proximate Information Technologies

Proximate education is currently and has been the predominant form of education for both the home and the school. In addition to a teacher (parents/siblings), our fictitious Susie has a workstation with a variety of media (book, paper, videotape) and a desk. Our proposed workstation utilizes existing media options (Figure 1). Current computer workstations are lesser approximations of what we can expect in the future.

The proximate technology and corresponding applications are currently further advanced than are the border zone or distant components. Impressive progress is being made with interactive videodisc and CD-ROM educational applications (proximate learning), but LAN, a vital component, is just being incorporated into educational settings, but not yet homes. These classroom/school level networks will be needed to internet with both the local wide area networks (border zone) and with national/international sources.

The microcomputer is the primary device that makes the model work. Currently, there are more than one million (probably close to 3,000,000 by 1989) microcomputers in K-12 classrooms, which is one for every 40 students (one for every 15 by 1989) (National Task Force on Educational Technology, 1986). Microcomputers in the home outnumbered school computers by a factor of 10 to 1; that is, one of every six families has a home computer (Komoski, 1984). With the exception of a computer coupled to a modem, there is no known home LAN programs in operation as such.

Classroom and/or school-wide LANs are slowly being added, and the most frequently installed is the Corvus Omni Net system (Birkhead, 1986). The most common LAN configurations are central office to school-wide classrooms; within a classroom, teachers are connected to 20-25 students. Most schools are purchasing add-on systems. There is currently no data available providing estimates concerning rate of LAN growth, but both public and private K-12 grade schools are purchasing networking systems (Birkhead, 1986; Epstein, 1985).

Border Zone Information Technologies

We believe that campus-wide LANs, which are rapidly being installed at most major colleges and universities, will be the backbone for border zone technological development. Carnegie-Mellon, MIT, and Brown Universities are but three of at least 20 major campus-wide network projects which serve as models (Chapman, 1986; Crecine, 1986). Campus networking is generally connected by either broadband or Ethernet systems which serve individuals, smaller LANS (departments), or campus-wide units (library) through interneted systems (Chapman, 1986). Some campus-wide LANs provide service for a 30-mile radius.

Because a central mission of universities is to create and disseminate information, universities are uniquely positioned to perform pivotal roles in border zone implementation. We think that border zone consortiums will be designed to link regional libraries, governmental computer centers (social, health, mental health agencies), hospitals, cablesystems, and telephone companies. These groups should participate in conceptualizing, designing, organizing, and operating the consortium enterprise. First-order services should provide regional citizens with network access to local and regional databases, an outgrowth of computerizing current record storage systems. The border zone consortium should develop an interneted "network of networks," ensure citizen equity and access, develop levels of access restrictions, and provide regional-wide access to distant technologies (satellite/cable two-way interaction, national/international hierarchy of networks).

Border zone development must include every social context in which public information is needed. Some believe that in the future, Susie

Anybody will receive a major portion of her education in the home (National Task Force on Educational Technology, 1986; Stephenson and deLandsheere, 1985). Although models of home border zone networks are few in number, we are beginning to see some innovative implementations. Columbus, Ohio, has a videotex system where citizens who view certain television programs may participate in two-way question and answer activities. Recently, the senior author had thirty families participate in a local two-way research project exploring family decision-making activities through networked home computers and a university main frame (Eichhorn, Watson, and Scanzoni, 1986). Home border zone applications are lagging behind public school, college, and university applications, but they should develop more rapidly as business and economics become involved (e.g., banking via home computer).

The border zone will be configured as a greatly expanded LAN. A large number of decisions will be necessary such as agreements concerning sets of conventions or communication protocols, interneting networks, establishing local/regional data bases, and development and management of the information infrastructure. The logistics will be challenging. One only has to imagine the task of organizing the massive amount of information (trillions of bytes) so that Susie can interact from her workstation with her teacher, her regional child development specialist, and the world.

Distant Information Technologies

Distant information technologies are generally new, but their roots are found in the telegraph, radio, telephone, film, and television. New distant technologies are television (especially two-way), cable, telephone, databases, networks, and satellites. Distant technology applications fit any social context (home as well as school). Unfortunately, because access currently is controlled through user cost, equity may become a major problem.

Distance technology sources are proliferating at a rapid rate. Over 3000 data base outlets are now available nationally (Lesko, 1984). Some of the better known data base sources Susie Anybody might want to use are: Source, CampuServe, BRS, and Dialog (Lesko, 1984; Lisanti, 1984). Wide-area networks are also developing rapidly. The NSFnet is a master plan designed to provide national networking for the entire academic and associated industrial research community, allowing an individual researcher access to fellow scholars anywhere in the nation. NSFnet will incorporate and interconnect all supercomputers, six wide-area networks (ARPANET, CSNET, BITNET, MFENET, UUCP, and USENET), state networks where available (Merit Computer Network in Michigan), the proposed NYSERNET, JVNC Princeton Consortium Center, SDSC (San Diego Consortium Center), and most campus networks (Jennings *et al.*, 1986). NSFnet should serve as a model for border zone development.

One finds a similar proliferation of satellites and satellite communication, television networks, cable systems, and telephone applications, which also produce unique opportunities for border zone exchange. We are truly headed toward global communication networks. Two groups—an alliance among IBM, CBS, and MCI and a separate effort by AT&T—are investing heavily in a global exchange of information by merging the computer and communications industries (Adams and Fuchs, 1986).

How will Susie Anybody adapt to distant education? She must utilize a higher level of self-involvement, since distant learning seems to be more individualized and user-controlled. Distant learning skills must be identified and taught to manage the new information infrastructure. Potential problems which could limit the future of distant learning technologies include equity of user access, societal acceptance of distant education, reconciliation of technological innovation, political and economic realities, and marketplace factors. The significant question will be whether and when the public decides that Susie is being educationally handicapped without access to a workstation.

Conclusion

We have proposed a proximate ↔ border zone ↔ distant educational model for integrating information technologies projected for the future. Our primary focus is on a learner workstation which we intend to examine in a series of research studies. Can a typical "Susie Anybody" successfully operate the proposed learner workstation? Are distant educational skills unique or distinct from those that have been successfully used in proximate learning? Much research and development is needed. We propose the following conclusions from our model building efforts:

1. A learner workstation is the information age equivalent of the desk.
2. Input/output controls will be word and graphics oriented with little need for programming languages.
3. Home and school workstations may be identical or altered to fit the specific needs of the person.
4. The workstation is the point at which proximate ↔ border zone ↔ distant education transpires.

5. Learning paradigms will shift from teacher oriented, norm-based approaches to teacher directed, individualized workstation applications.
6. Proximate and distant education will move toward learning packets predicated on cognitive styles, levels of intrinsic motivation, self-pacing, learner control, and learner management.
7. College/university professors' descriptive metaphors will continue to change from teacher and knowledge repositors to creators and managers of the information infrastructure.
8. Public school teachers' metaphors also will continue to change from "teachers" and keepers of the subject matter knowledge to learning directors, coaches, contractors, and diagnosticians. Teachers will have to share the educational enterprise with many new partners in the proximate ↔ border zone ↔ distant domains.

Finally, we expect that the educational system with which Susie Anybody interacts in the year 2000 will integrate home, school, and other social contexts. By this time there should be widespread acceptance of a shift in information repositories from the human mind to mechanical storage devices (CD-ROM, diskettes, videodiscs, videotapes). As we accept the magnitude of the knowledge explosion, information management will replace retention of facts as the primary student goal. We also predict that learning curricula will shift toward teaching higher order thinking. We will seek to train our "Susies" to be better at solving problems, modeling reality, constructing and testing theory, enumerating variable interactions, investigating alternate viewpoints, and constructing symbolic hierarchies. By the year 2000, public schools, colleges, and universities will integrate distant education with today's proximate education through the use of technological innovations like the workstation. □

References

Adams, D.M., and Fuchs, M. Toward Global Communication Networks: How Television Is Forging New Thinking Patterns. *T.H.E. Journal*, 1986, *13*(5), 108-111.

Birkhead, E. Technology Update: Installed Base of LANs Will Skyrocket During 1986-87 School Year. *T.H.E. Journal*, 1986, *13*(9), 12-13.

Chapman, D.T. Campus-wide Networks: Three State-of-the-Art Demonstration Projects. *T.H.E. Journal*, 1986, *13*(9), 66-70.

Crecine, J.P. The Next Generation of Personal Computers. *Science*, 1986, *231*(4741), 942-953.

Eichhorn, M.F., Watson, J.A., and Scanzoni, J. The Use of Microcomputers in Family Studies. Paper presented at the National Council of Family Relations, Dearborn, Michigan, November, 1986.

Epstein, J.L. Home and School Connections in Schools of the Future: Implications of Research on Parent Involvement. *Peabody Journal of Education*, 1985, *62*(2), 18-41.

Jennings, D.M., Landweber, L.H., Fuchs, I.H., Farber, D.J., and Adrion, W.R. Computer Networking for Scientists. *Science*, 1986, *231*(4741), 943-950.

Komoski, P.K. Educational Computing: The Burden of Insuring Quality. *Phi Delta Kappan*, 1984, *66*, 244-248.

Lesko, M. Low-Cost On-line Databases. *Byte*, 1984, *9*(11), 167-176.

Lisanti, S. The On-line Search. *Byte*, 1984, *9*(13), 215-230.

National Task Force on Educational Technology. Transforming American Education: Reducing the Risk to the Nation. (A report to the Secretary of Education, United States Department of Education). *T.H.E. Journal*, 1986, *14*(1), 58-67.

Pea, R.D. Beyond Amplification: Using the Computer to Reorganize Mental Functioning. *Educational Psychology*, 1985, *20*(4), 167-182.

Stephenson, B., and deLandsheere, G. Excerpts from the International Conference on Education and New Information Technologies. *Peabody Journal of Education*, 1985, *62*(2), 75-92.

Watson, J.A., Calvert, S.L., and Brinkley, V.M. The Computer/Information Technologies Revolution: Controversial Attitudes and Software Bottlenecks—A Mostly Promising Progress Report. *Educational Technology*, in press, 1987.

Article Citations

1. The New Age of Telecommunication: Setting the Stage for Education. By Dan J. Wedemeyer. *Educational Technology*, October 1986, pages 7-13.
2. Educational Technology Use in Distance Education: Historical Review and Future Trends. By Robert A. Gray. *Educational Technology*, May 1988, pages 38-42.
3. Building Connections for the Growth of Distance Education. By Colin W. Dunnett. *Educational Technology*, December 1984, pages 30-31.
4. Telecommunications and the Building of Knowledge Networks: Here Today, Much More Tomorrow. By Dennis Adams and Mary Hamm. *Educational Technology*, September 1988, pages 51-53.
5. Linking Teachers to the World of Technology. By Gerald Marker and Lee Ehman. *Educational Technology*, March 1989, pages 26-30.
6. Literacy in the Electronic Age. By Peter H. Wagschal. *Educational Technology*, June 1987, pages 5-9.
7. Communications Satellites: A Rural Response to the Tyranny of Distance. By Gregory Jordahl. *Educational Technology*, February 1989, pages 34-38.
8. Telecommunications in the Classroom: Can It Be Done? Should It Be Done? An Essay on Possibilities and Frustrations. By David W. Swift. *Educational Technology*, May 1983, pages 23-25.
9. Electronic Mail: An Exemplar of Computer Use in Education. By Heinz V. Dreher. *Educational Technology*, August 1984, pages 36-38.
10. Tailoring Telecommunications Innovations to Fit Educational Environments. By Judith B. Harris. *Educational Technology*, November 1989, pages 7-11.
11. Computer Mediated Communication for Instruction: Using e-Mail as a Seminar. By Alexander J. Romiszowski and Johan A. de Haas. *Educational Technology*, October 1989, pages 7-14.
12. Teleconferencing: An Instructional Tool. By Resa Azarmsa. *Educational Technology*, December 1987, pages 28-32.
13. Computer Conferencing: Models and Proposals. By Valarie Meliotes Arms. *Educational Technology*, March 1988, pages 43-45.
14. Learning Over the Lines: Audio-Graphic Teleconferencing Comes of Age. By Michael K. Gardner, Sidney Rudloph, and Gabriel Della-Piana. *Educational Technology*, April 1987, pages 39-42.
15. Use of Audiographic Technology in Distance Training of Practicing Teachers. By Dennis R. Knapczyk. *Educational Technology*, June 1990, pages 24-27.
16. The Design and Application of a Distance Education System Using Teleconferencing and Computer Graphics. By Bill Winn, Barry Ellis, Emma Plattor, Larry Sinkey, and Geoffrey Potter. *Educational Technology*, January 1986, pages 19-23.
17. Teletext: A Distance Education Medium. By Aliza Duby. *Educational Technology*, April 1988, pages 54-57.
18. Local and Long Distance Computer Networking for Science Classrooms. By Denis Newman. *Educational Technology*, June 1987, pages 20-23.
19. Computer CONFERencing in the English Composition Classroom. By Rosemary Kowalski. *Educational Technology*, April 1989, pages 29-32.

20. Busy Professionals Go to Class the Modem Way: A New Approach to Distance Learning in the Electronic Classroom. By Michael Thombs, Patricia Sails, and Beverly Alcott. *Educational Technology*, October 1989, pages 30-31.

21. AIDS Training in Third-World Countries: An Evaluation of Telecommunications Technology. By Pamela D. Hartigan and Ronald K. St. John. *Educational Technology*, October 1989, pages 20-23.

22. Inservice Training Via Telecommunications: Out of the Workshop and Into the Classroom. By Joseph J. Stowitschek, Brent Mangus, and Sarah Rule. *Educational Technology*, August 1986, pages 28-33.

23. Extending Education Using Video: Lessons Learned. By Dean R. Spitzer, Jeanne Bauwens, and Sue Quast. *Educational Technology*, May 1989, pages 28-30.

24. Some Advantages and Disadvantages of Narrow-Cast Television: One Instructor's Experience. By David W. Dalton. *Educational Technology*, January 1987, pages 42-44.

25. An Apple a Day and at Night: A Distance Tutoring Program for At-Risk Students. By Steven M. Ross, Lana Smith, Gary Morrison, and Ann Erickson. *Educational Technology*, August 1989, pages 23-28.

26. Telecommunications Skills Training for Students with Learning Disabilities: An Exploratory Study. By Barbara J. Edwards and Mark A. Koorland. *Educational Technology*, January 1990, pages 34-36.

27. Live from Germany: A Foreign Language Encounter Via Satellite. By Hildburg Herbst and Peter Wiesner. *Educational Technology*, April 1988, pages 41-43.

28. KITES: A Middle School Environmental Science Telelink to West Germany. By John LeBaron and Virginia Teichmann. *Educational Technology*, December 1989, pages 51-53.

29. Critical Barriers to the Adoption of Instructional Television in Higher Education. By F.R. (Bud) Koontz. *Educational Technology*, April 1989, pages 45-48.

30. The Use of Interactive Television in Business Education. By Gordon D. Pirrong and William C. Lathen. *Educational Technology*, May 1990, pages 49-54.

31. The Independent/Distance Study Course Development Team. By Clayton R. Wright. *Educational Technology*, December 1988, pages 12-17.

32. How Television Is Changing Thinking Patterns as We Move Toward Computer Controlled Global Communication Networks. By Dennis M. Adams and Mary Fuchs. *Educational Technology*, March 1986, pages 26-29.

33. Computing and Telecommunications in Higher Education: A Personal View. By Norman Coombs. *Educational Technology*, February 1990, pages 46-47.

34. Technology: Implications for Long-Range Planning. By David Foster. *Educational Technology*, April 1988, pages 7-14.

35. Why Information Technologies Fail. By Diane Gayeski. *Educational Technology*, February 1989, pages 9-17.

36. Human and Quality Considerations in High-Tech Education. By Peter Smith and Samuel Dunn. *Educational Technology*, February 1987, pages 35-39.

37. Distance-Delivered Instruction: Making It Work. By Barry Willis. *Educational Technology*, July 1989, pages 46-47.

38. Distance Education and Student Procrastination. By Thomas W. Wilkinson and Thomas M. Sherman. *Educational Technology*, December 1989, pages 24-27.

39. Distance Learning and Public School Finance. By Brian D. Monahan and Charles Wimber. *Educational Technology*, July 1988, pages 41-43.

40. An Information Technologies Workstation for Schools and Homes: Proximate, Border Zone, and Distant Educational Possibilities for the Future. By J. Allen Watson, Sandra L. Calvert, and Roseann Collins. *Educational Technology*, November 1987, pages 14-20.

Index